CONSTRAINT SOLVING OVER MULTI-VALUED LOGICS

Frontiers in Artificial Intelligence and Applications

Volume 91

Published in the subseries

Dissertations in Artificial Intelligence

Under the Editorship of the ECCAI Dissertation Board
Proposing Board Member: Luís Moniz Pereira

Recently published in this series:

ISSN: 0922-6389

Constraint Solving over Multi-valued Logics

Application to Digital Circuits

Francisco Azevedo

Departamento de Informática, UNL/FCT, Caparica, Portugal

IOS

Press

Ohmsha

Amsterdam • Berlin • Oxford • Tokyo • Washington, DC

ISBN 1 58603 304 2 (IOS Press)
ISBN 4 274 90555 1 C3055 (Ohmsha)
Library of Congress Control Number: 2002113947

Publisher
IOS Press
Nieuwe Hemweg 6B
1013 BG Amsterdam
The Netherlands
fax: +31 20 620 3419
e-mail: order@iospress.nl

Distributor in the UK and Ireland
IOS Press/Lavis Marketing
73 Lime Walk
Headington
Oxford OX3 7AD
England
fax: +44 1865 75 0079

Distributor in the USA and Canada
IOS Press, Inc.
5795-G Burke Centre Parkway
Burke, VA 22015
USA
fax: +1 703 323 3668
e-mail: iosbooks@iospress.com

Distributor in Germany, Austria and Switzerland
IOS Press/LSL.de
Gerichtsweg 28
D-04103 Leipzig
Germany
fax: +49 341 995 4255

Distributor in Japan
Ohmsha, Ltd.
3-1 Kanda Nishiki-cho
Chiyoda-ku, Tokyo 101-8460
Japan
fax: +81 3 3233 2426

ACKNOWLEDGMENTS

First of all, I would like to thank my family, my parents and my sister for all the help, support and, specially, love they provided me during my whole life and through the hard but stimulating times of my Ph.D. studies with all their ups and downs. To them, I owe my education and my longing for Truth, wisdom, knowledge and intelligence (even if artificial). I also remember my late grandparents with all their care.

As to this research work, I had the honour of having Pedro Barahona as my supervisor and friend. I cannot thank him enough for all the patience in reading, understanding and proposing improvements to my writings, amidst all his works. He introduced and pushed me to this field where he has taught me so much.

Thank you to Paulo Flores and his supervisor João Paulo Marques Silva of the INESC team, for getting me acquainted with the problems of the ECAD area, and for sending me precious data and tools even in troubled times.

I also would like to thank the IC-Parc team, especially Mark Wallace, Carmen Gervet, and Joachim Schimpf, for the warm welcome, hospitality, friendship, and fruitful discussions during my stay there, which reflected in this dissertation.

My acknowledgements to the Department of Computer Science and CENTRIA for providing all needed equipment and working conditions, to my colleagues such as Carlos Damásio, João Leite, António Ribeiro, Rui Marques or Jorge Cruz for the help and for being there, and to the secretary staff for the support and general help.

Naturally, I thank the financial support of "Sub-Programa Ciência e Tecnologia do 2º Quadro Comunitário de Apoio" that always arrived on time. In addition, FACC ("Fundo de Apoio à Comunidade Científica") of FCT ("Fundação para a Ciência e Tecnologia") from Portugal, kindly granted me a subsidy that made the printing of this book possible, thanks to "Apoio do Programa Operacional Ciência, Tecnologia, Inovação do Quadro Comunitário de Apoio III".

And, of course, thanks be to God, for without Him no thanks would be possible!

FCT Fundação para a Ciência e a Tecnologia
MINISTÉRIO DA CIÊNCIA E DO ENSINO SUPERIOR   Portugal

Universidade Nova de Lisboa
Faculdade de Ciências e Tecnologia

Abstract

CONSTRAINT SOLVING OVER MULTI-
VALUED LOGICS — APPLICATION TO
DIGITAL CIRCUITS

by Francisco de Moura e Castro Ascensão de Azevedo
Supervisor: Professor Pedro M. C. C. Barahona
Departamento de Informática

Due to usage conditions, hazardous environments or intentional causes, physical and virtual systems are subject to faults in their components, which may affect their overall behaviour. In a '*black-box*' system modelled by a set of propositional logic rules, in which just a subset of components is externally visible, such faults may only be recognised by examining some output function of the system. A (fault-free) model of the system provides the expected output given some input. If the real output differs from that predicted output, then the system is faulty. However, some faults may only become apparent in the system output when appropriate inputs are given. A number of problems regarding both testing and diagnosis thus arise, such as testing a fault, testing the whole system, finding possible faults and differentiating them to locate the correct one. The corresponding optimisation problems of finding solutions that require minimum resources are also very relevant in industry, as is minimal diagnosis.

In this dissertation we use a well established set of benchmark circuits to address such diagnostic related problems and propose and develop models with different logics that we formalise and generalise as much as possible. We also prove that all techniques generalise to multiple faults and to systems modelled by a set of propositional logic rules. The developed multi-valued logics extend the usual Boolean logic (suitable for fault-free models) by encoding values with some dependency (usually on faults). Such logics thus allow modelling an arbitrary number of diagnostic theories. Each problem is subsequently solved with CLP solvers that we implement and discuss, together with a new efficient search technique that we present. We compare our results with other approaches such as SAT (that require substantial duplication of circuits), showing the effectiveness of constraints over multi-valued logics, and also the adequacy of a general set constraint solver (with special inferences over set functions such as cardinality) on other problems. In addition, for an optimisation problem, we integrate local search with a constructive approach (branch-and-bound) using a variety of logics to improve an existing efficient tool based on SAT and ILP.

GLOSSARY OF ACRONYMS

AC	Arc Consistency
AI	Artificial Intelligence
ATG	Automatic Test Generation
ATMS	Assumption-Based TMS
BAC	Bounded Arc Consistency
BB	Branch and Bound
BIST	Built-In Self-Test
Bit	Binary Digit
CLP	Constraint Logic Programming
CNF	Conjunctive Normal Form
CSP	Constraint Satisfaction Problem
CP	Constraint Programming
CPU	Central Processing Unit
CUT	Circuit Under Test
DAC	Directional Arc Consistency
DTG	Differential Test Generation
ECAD	Electronic Computer-Aided Design
FC	Forward Checking
FD	Finite Domain
FF	First-Fail
GHz	Giga Hertz
HCLP	Hierarchical CLP
I/O	Input/Output
ILP	Integer Linear Programming
ISCAS	International Symposium on Circuits and Applied Systems
ITBS	Iterative Time-Bounded Search
LP	Logic Programming
LS	Local Search
MAC	Maintaining Arc Consistency
Mb	Mega bytes
MHz	Mega Hertz
MSF	Multiple Stuck Fault
MTP	Minimum Test Pattern
PC	Path Consistency
PI	Primary Input
PO	Primary Output
RAM	Random Access Memory
RISC	Reduced Instruction-Set Computer
s-a-0	Stuck at 0
s-a-1	Stuck at 1
SAT	Propositional Satisfiability
SSF	Single Stuck Fault
TG	Test Generation
TMS	Truth Maintenance System
VLSI	Very Large Scale Integration
XOR	Exclusive OR

TABLE OF CONTENTS

LIST OF FIGURES

LIST OF TABLES

Introduction

Due to manufacturing errors, long-usage, hazardous environment, or some other cause, physical systems are subject to *faults* in their components, which may affect the overall system behaviour. In a system modelled by a set of *propositional rules*, but with just a subset of components externally visible (thus behaving like a *'black-box'*), such faults may only be recognised by examining some *output function* of the system. By knowing the *input* given to the system, its initial (fault-free) *model* provides the expected output. If the real output differs from that predicted output, then the system is faulty. However, some faults may only become apparent in the system output under certain conditions, i.e. when appropriate inputs are given. This fact poses a number of problems regarding both the **testing** and **diagnosis** of systems. For example (to name just a few):

- How to test a system? What is the smallest amount of effort required for that?
- How to test a fault? What are the minimum resources required ?
- What are the possible faults ?
- How to differentiate between possible faults and locate the correct one ?

The systems of interest to this dissertation are, in general, logical-based 'black-box' systems modelled by a set of propositional rules, with some controllable inputs and some visible outputs, all Boolean (some more complex agents can be transformed into such systems). An output bit may unveil a fault if, when given some input to the system, its value is dependent on the presence (or absence) of the fault. Such fault is not restricted to be a physical fault since it can also be applied to virtual systems (e.g. on the internet). Virtual systems may also have a known (or expected) model and behave as faulty due to some (possibly intentional) change of belief or some relaxed rule, for instance. This is particularly relevant in the emerging field of electronic transactions in a society of agents, where rules for bids in an auction depend on knowledge acquired on competitors and their actions. It is thus very important to predict the behaviour of the agent with whom some contract is being negotiated, after some unexpected output is detected. Hence, it is crucial to test the agent, formulate hypotheses for its unexpected behaviour, refine them in order to obtain a minimal set of hypotheses (possibly by performing additional tests), and model them. Having modelled the competitors, an optimal 'bid' (the one that is expected to generate the best 'price') can be found.

A number of problems concerning such systems have been deserving especial attention by the scientific community. In particular, digital circuits in the Electronics Computer Aided Design (ECAD) area have traditionally established several problems, for which widely used benchmarks are available. The use of digital circuits has known a great expansion for many years and they are now present virtually everywhere. Reflecting the economic importance of this area, such problems have been widely studied and they are still the subject of active research, with a variety of approaches. The evolution of the area, with new technologies and continuous new requirements and needs, makes it a suitable application field for Constraint Logic Programming (**CLP**) [Jaffar and Lassez 1987, Jaffar and Maher 1994], which combines the declarativity of logic programming with the power and efficiency of constraint reasoning (see section 1.3). Such suitability was already exemplified and discussed in this area [Simonis 1992] and we reckon that the expressive power and flexibility of the constraint programming approach makes it very attractive to tackle additional problems over circuits in particular, and logical-based systems in general.

The traditional Boolean domain and corresponding logic is sufficient to model a normal digital circuit, by representing each gate as a Boolean operation. However, since physical circuits may be faulty, such model needs to be extended somehow in order to model circuits that may exhibit a different behaviour than that predicted by a model of the normal circuit. While this may be simple when all gates are accessible, it is not so when the circuit is, in practice, a 'black-box' in which only a set of inputs and a set of outputs are visible. Since it is not known *a priori* whether a gate is normal or faulty, such fact may be modelled as a disjunction to represent the alternative choices. Such simple modelling has however the drawback of possibly leading to excessive backtracking in problem solving.

The use of extra symbolic values in conjunction with the usual Boolean $\{0,1\}$ was proposed in a D-calculus [Roth 1966] to reflect the dependency of values in physical lines on some faulty gate. When the gate is stuck at some Boolean value v (i.e. its output is always v, regardless of the inputs), the physical value of its output depends in general on whether it is really faulty or not. Moreover, gates that take that value as input can also be affected, and hence propagate the fault effect. Two extra values are then used in the model of the circuit to denote that the normal (expected) value 0 or 1 in some line is inverted if the fault is present. Hence, multi-valued logics are introduced to model circuits in which signal values may depend on faults affecting the circuit. In this case, a 4-valued logic is defined to model dependencies on a single fault. Gate operations involve four values and their semantics have to be defined by extended 'truth' tables that, implicitly, consider two (normal and faulty) circuits.

The use of multi-valued logics is not new in computer science, with successful application in several domains such as temporal logics [Prior 1957, Galton 1987, van Benthem 1995] or in qualitative reasoning [Forbus 1984, Davis 1984, Kuipers 1989]. Modelling dependencies with such logics relates to *truth maintenance systems* (TMSs) [Doyle 1979], as discussed in section 1.2, and has possible applications for generic systems (and for circuits in particular) for tasks such as diagnosis or general modelling to predict behaviour.

The introduction of multi-valued logics induces problem variables (signal lines) with finite domains, thus suggesting the use of CLP(FD) for efficient solving. Instead, separate circuit models with Boolean variables can be considered and subsequently solved by propositional satisfiability (SAT) approaches [Marques-Silva 1995].

In this dissertation we address a number of diagnostic related problems (in the sense that faults are taken into account) and propose and develop models with different logics that we formalise and generalise as much as possible. In addition, we try to solve each problem with CLP solvers that we implement and discuss, comparing with other approaches. The adequacy of the proposed approaches is illustrated with examples and results, pointing to other applications (where appropriate) and possible research directions.

1.1 Scope

This dissertation covers a number of satisfaction and optimisation problems, such as modelling of circuits and faults, simulation of normal and faulty behaviour, test generation, minimal and differential diagnosis, generation and minimisation of test sets, unveiling and maximising faults detected by a test, and maximisation of unspecified inputs of a test. Techniques developed apply to these and other related problems over circuits in particular and systems in general.

While such problems concern general 'black-box' systems modelled as propositional theories, examples throughout this dissertation consider only the particular but representative case of combinational digital circuits, for which a well established set of benchmarks [ISCAS 1985] exist, thus allowing comprehensive comparisons of results with different approaches for those problems. Applicability of developed techniques to general systems is trivially proven in the last chapter.

Some examples and notation concerning circuits are taken from [Abramovici *et al.* 1990], where a thourough study of digital systems testing can be found and where it is also shown that the consideration of combinational circuits alone is not too restrictive, since problems over sequential circuits are usually best handled with techniques developed for combinational circuits (often by transforming those circuits into combinational).

The types of faults considered are the usual stuck-at faults, where a gate is stuck at some Boolean value v thus always outputting v regardless of its input. Again, [Abramovici *et al.* 1990] shows that such faults are not too restrictive since other types of faults can often be considered through these simpler stuck-at faults, and such (apparent) restriction allows for more efficient techniques to be developed. Furthermore, when talking about virtual systems modelled as a set of propositional rules, these faults are the only ones that need to be considered.

1.2 Truth Maintenance Systems

Truth maintenance systems (TMSs) [Doyle 1979] reason about logical statements in order to maintain global consistency. Inferences are justified by a set of supporting statements that are recorded as dependencies. Such systems support nonmonotonic reasoning since assumptions derived from default reasoning may be revised due to contradictory new information (inferred statements). For that, TMSs refer to a statement as a node that may be at any point in either of two states: IN or OUT. If a node is believed to be true then it is IN, otherwise it is OUT either because there is no justification for it to be true or because justifications for it are currently not valid.

Each node n then has a list of justifications attached, which basically are support lists mentioning nodes that must be IN and nodes that must be OUT for n to be believed to be true (IN).

When a *contradiction* arises, a dependency-directed backtracking procedure is invoked on the contradiction node created, looking at its justification nodes and recursively at their justifications until a set of inconsistent assumptions is found, which is then recorded as a *nogood* node to avoid further computations based on it.

To model circuits problems, a digital gate is assumed to be either normal or with a specific fault. Inspection of the output may then reveal a contradiction that leads to a set of inconsistent assumptions, which are however found only indirectly since dependency on assumptions is not explicitly expressed (in fact, the system has to search recursively through justifications to find such dependencies).

In conventional *justification-based* TMSs, a particular datum is believed in an implicit context given by the set of all IN nodes. Context switching is performed for truth maintenance by dependency-directed backtracking to move inside the search space. To avoid most backtracking and easily switch between contexts, Johan de Kleer proposed [1986] an *assumption-based* TMS (ATMS), where data are labelled with the sets of assumptions (representing contexts) under which they hold. Such sets are kept as general as possible in order to more easily derive data (for example, if context C includes context B, then all derivations of B remain valid under C). Similarly, contradictions are recorded in the most general form to rule out as much of the search space as possible. Thus, multiple solutions can be explored simultaneously without the need for the global database to be consistent. For this, ATMSs spend most computation time in determining most general forms.

While conventional TMSs are oriented towards finding one solution, the ATMS is oriented to finding all solutions. Hence, ATMSs are best suited only for problems with many solutions and where all of them are required.

The multi-valued logics that we develop in this dissertation allow more efficient tools to be developed by encoding direct dependency on the assumptions that care for each problem. Hence, overhead of searching dependencies with TMS is avoided. And since, in general, solving 'circuit' problems requires finding a solution rather than all of them, overhead of considering all contexts as in ATMSs is also avoided. Furthermore, for such cases where all solutions are required, such as in diagnosis, it is often possible to be more efficient using a multi-valued logic as reported in [Alferes *et al.* 2001], where a model using a logic over Boolean and set values generates all possible diagnoses (for an incorrect circuit behaviour) in a single step with no backtracking at all!

1.3 Constraint Reasoning

A problem where relations between variables (to which values must be assigned) are restricted by a number of constraints that must be satisfied, is referred to as a Constraint Satisfaction Problem (CSP) [Montanari 1974, Mackworth 1977]. The goal is to assign values to all the variables without violating any constraint, or to prove this to be impossible. The space formed by all possible combinations of assignments is referred to as the search space. Such combinatorial search problems have been widely studied in *Artificial Intelligence* (AI) and *Operations Research*.

CLP(*FD*) integrates constraints in logic programming, where variables may be restricted to explicitly range over finite subsets of the universe of values, thus defining finite domain variables [Van Hentenryck and Dincbas 1986]. Search for solutions involve assigning variables with values from their domain and, when a contradiction is found due to some violated constraint, perform some form of backtracking [Golomb and Baumert 1965] (usually chronological) to undo choices made and try other yet unexplored branch of the search tree.

Since the search space of combinatorial problems is usually intractable [Garey and Johnson 1979], a naïve *generate-and-test* approach, in which each combination of possible assignments is generated and then tested for a solution, until one is found, is unsuitable. Hence, *constraint reasoning* techniques are usually applied in AI to (often, drastically) reduce search space by discarding impossible solutions [Mackworth 1977, Nadel 1989, Dincbas *et al.* 1990, Dechter 1992, Mackworth 1992, Kumar 1992]. In this section, we briefly describe the basics of such local consistency techniques that look ahead at logical predicates defined as constraints to discard impossible values from the domain of individual variables.

1.3.1 Consistency Techniques

Constraint solvers apply *constraint propagation* or *consistency techniques* [Van Hentenryck 1989] in order to remove redundant (i.e. impossible) values from the domains of variables involved in stored constraints. If the domain of some variable becomes empty after the application of such techniques, then the CSP is insoluble. Otherwise, the CSP is said to be **consistent** (with regard to some properties) and there *may be* a solution, which has to be found to definitely prove one exists. In general, solvers using such techniques are incomplete, which means that reaching a *consistent* state for the CSP is not a sufficient condition for its satisfiability. Hence, a search phase must still occur to find a possible assignment of values to all the variables. Consistency techniques are interleaved during this search phase to constantly reduce search space, aiming at saving computation time. Notice that there is a trade-off between the level of consistency applied and the amount of pruning (of the search tree) obtained. Usually, greater levels of consistency produce decreasingly larger improvements on pruning and require increasingly larger amounts of CPU time, thus becoming counter-productive at some stage.

In this section we briefly describe the most common consistency techniques, each guaranteeing a different property. Definitions are taken from [Tsang 1993], where a thorough study on these concepts can be found.

Before introducing such concepts we first recall that a CSP is composed of a set of variables (each with some domain) related by a set of constraints that restrict the values they can simultaneously take. A **binary** CSP is one such problem where constraints are just unary or binary (i.e. each constraint involves just one or two variables). It can be represented as a graph where nodes are variables and edges are constraints relating two nodes. A CSP not limited to such constraints is referred to as a **general** CSP and defines a hypergraph where hyperedges are constraints relating (connecting) an arbitrary number of variables (nodes). Notice, however, that any CSP has a **dual** binary CSP (as described in [Tsang 1993]).

To solve a CSP, variables have to be assigned values from their domains (i.e. labelled). An assignment of value v to variable x is denoted by a **label** in the form $<x,v>$. A **compound label** is a set of labels representing a simultaneous assignment of values to different variables.

With these notions we may now define some consistency techniques.

A CSP is **node-consistent** (NC) if and only if for all variables all values in its domain satisfy the constraints on that variable (i.e. unary constraints).

Node consistency is usually applied in CLP(*FD*) since the finite enumeration of possible values allows an efficient removal of values that do not satisfy a constraint.

A CSP is **arc-consistent** if and only if for every variable x, for every label $<x,a>$ that satisfies the constraints on x, there exists a value b for every variable y such that the compound label ($<x,a> <y,b>$) satisfies all the constraints on x and y.

For n-ary constraints, *generalised arc consistency* can also be defined and applied by checking compatible labels among all variables related by each constraint. However, this is generally too expensive.

To keep a CSP arc-consistent there are a number of algorithms that go through the variables to remove incompatible values (with some other variable) from their domains. These algorithms vary on the number of steps and tests required to achieve AC, i.e. on their time and space complexity. We may refer, for binary constraints, the naïve AC-1 algorithm, which considers all constraints and for each value removed from a domain, it will go through all constraints again to check if some other value can be removed. AC-1 was then improved to AC-2 and AC-3, which would add the notion of supporting values to avoid checking all constraints again, but rather only those constraints that could be affected by the removal of some value from the domain of a variable. These algorithms were later further improved by the (considered optimal) AC-4, which extends the notion of support to conclude that although removing a value from the domain of x may affect y, the particular removal of v does not. Nevertheless, other algorithms were still developed [Bessière and Cordier 1994, Bessière *et al.* 1995] which can behave better in some situations.

Beyond AC one may still define **Path Consistency** (PC):

A CSP is path-consistent if and only if for all variables x and y, whenever a compound label ($<x,a> <y,b>$) satisfies the constraints on both x and y, there exists a label $<z,c>$ for every variable z such that ($<x,a> <y,b> <z,c>$) satisfies all the constraints on x, y and z.

Path consistency (or any variant or simplification) is rarely used due to its computational complexity. In practice, such kind of consistency may only be considered for binary CSPs or when constraints are few and domains are very small.

Since full AC is, in general, already too costly computationally, some consistency techniques try to approximate it by relaxing some properties. This implies that values incompatible with other variables may remain in some domain, but often it is worth that 'risk' since the relaxed technique is much simpler than full AC, and the pruning achieved is similar. Below we describe some of these techniques.

If there is a total ordering of variables, consistency techniques such as AC (and PC) can be directional. **Directional Arc-Consistency** (DAC) [Dechter and Pearl 1988] is thus defined when the constraint graph is in fact a tree:

> A CSP is *directional arc-consistent* under an ordering of the variables if and only if for every label $<x,a>$ that satisfies the constraints on x, there exists a compatible label $<y,b>$ for every variable y that is <u>after</u> x according to the ordering.

A weaker but often more efficient consistency technique than AC is that of **Bounded Arc-Consistency** (BAC) [Van Hentenryck 1989] in which only the two bounds of each variable (domain) are verified and possibly updated (tightened) to remove impossible values between two variables by restricting their domains. This is particularly used in CLP(*Intervals*), often for modelling domains over real numbers or sets, for example. In such cases, domains are expressed by a range in the form *lower-bound..upper-bound*.

1.3.2 Maintaining Consistency

Forward Checking (FC) is a search strategy that assumes node consistency and approximates arc-consistency by removing from domains of variables, those values that are incompatible with each assignment. I.e. whenever there is a commitment to a label $<x,a>$, FC removes all values from the domains of variables (other than x) that are incompatible with value a for x. If some domain becomes empty, unsatisfiability for the current set of assignments is detected and backtracking is forced. **Constraint propagation** is thus performed on instantiation of some variable. Arc consistency is forced between two variables only when one of them is labelled, thus being effectively reduced to node consistency.

The algorithm of **Maintaining Arc Consistency**, MAC [Sabin and Freuder 1994], is a combination of FC and AC in that the constraint network is made arc consistent initially and whenever there is a commitment to a label $<x,a>$, the effects of removing values (other than a) from the domain of x are propagated through the constraint network as necessary to restore full arc consistency. The particular usefulness of MAC in random binary CSPs is shown in [Sabin and Freuder 1994, Bessière and Régin 1996].

1.3.3 Advanced Techniques

In this section we mention different advanced techniques that CLP solvers may also apply. Such techniques will be referred along this dissertation to solve different problems.

The techniques described above use a local consistency criterion to remove impossible values from domains of variables by examining directly a constraint that relates them. However, often a set of basic constraints relating a set of variables can be seen as a **global constraint** [Beldiceanu 1990] over them, which allows specialised information to be used for further pruning of the search tree. For instance, let us assume that four variables X_1, X_2, X_3, X_4, with domain $\{1,2,3\}$, are constrained to be pairwise different (i.e. $X_1 \neq X_2 \wedge X_1 \neq X_3 \wedge X_1 \neq X_4 \wedge X_2 \neq X_3 \wedge X_2 \neq X_4 \wedge X_3 \neq X_4$), to express that all four variables are different. Clearly, there is no solution that satisfies those constraints since there are only 3 possible values for 4 different variables. Nevertheless, it can be verified that the CSP is path-consistent, since for any assignment to two variables, there is still one possible value left for a third variable. For such cases, a global constraint such as *all_different*($\{X_1, X_2, X_3, X_4\}$) may be used to decide for unsatisfiability due to shortage of resources (number of different values to assign). An efficient algorithm for such constraint is described in [Régin 1994].

A global constraint may be either user-defined or provided as built-in of a CLP solver (a number of such built-in global constraints exist, e.g. [Beldiceanu and Contejean 1994]).

Similarly, a *meta-constraint* such as the *cardinality* operator [Van Hentenryck and Deville 1991] may be used for a more global reasoning. The cardinality constraint is given a set of goals, of which at least n and at most m must be satisfied. It may thus represent a disjunctive constraint, which allows delaying choices (the disjuncts) by reasoning globally on the disjunction to, when some goals become known to be impossible or trivially satisfied (during search or upon new constraints posted), infer that the remaining goals must be either true or false.

A yet more powerful technique to handle disjunctions consists of *constructive disjunction* [le Provost and Wallace 1993] which may restrict domains of variables by reasoning globally on the disjunction (enforcing AC).

In general, interleaved with the maintenance of some form of consistency, a *labelling* phase for enumeration of the CSP variables must still occur to prove whether a solution exists and find it. There are different ways to go through the search space, with possible dramatic differences in efficiency. Hence, in addition to constraint propagation techniques, it is also important to pay attention on guiding search.

Instantiating variables with values from their domains sequentially in a depth-first search, with chronological backtracking of Prolog, is often unsatisfactory. **Heuristics** (functions to rank candidates) may thus be used to choose what variable to instantiate next and what value should be tried first. The difference in ordering may be critical since there may be labels much more likely to belong to a solution than others, and because some variables may be harder to instantiate. This is the case considered by the popular *first-fail* (FF) heuristic [Haralick and Elliott 1980], which selects variables with smaller domains first.

To avoid being "stuck" at an incorrect early choice of a depth-first search, *randomisation* techniques may also be applied to "jump" to a different point in the search space. This allows other branches to be explored before some defined threshold is reached, which contributes to a "fairer" distribution of search and, consequently, to increase probability of finding a solution faster. Existing such techniques include *iterative broadening* [Ginsberg and Harvey 1990], *iterative deepening* [Meseguer 1997], and *limited discrepancy search* [Harvey and Ginsberg 1995, Walsh 1997]. A new search technique based on randomisation is presented in Chapter 3 and is subsequently applied to solve more complex problems addressed in following chapters.

For an optimisation goal, one is not interested just in finding a solution but rather in finding the best one. For that, once some solution is found, only better solutions (according to some measure function) are sought. Hence, partial solutions already known to have no possibilities of improving the current optimal solution can be discarded, and other branches of the remaining search tree can be explored. Such pruning and branching technique according to a current bound is referred to as **branch-and-bound** [Papadimitriou and Steiglitz 1982, Balas and Toth 1985].

Often, a solution found can be improved more efficiently by performing some form of **local search**, in which small changes to the current solution are explored to find a local optimum.

In Chapter 7 we combine branch-and-bound and local search techniques to solve an optimisation problem using models with different multi-valued logics.

1.4 Contributions and Limitations

The contributions of this dissertation are not restricted to a single domain bur rather aim at different aspects of computer science in general, and AI in particular. The areas potentially benefited involve *Constraint Reasoning*, formal *Multi-valued Logics*, ECAD *problem modelling and solving* (namely simulation, test generation, optimisation, basic and differential diagnosis) and *Agents* (since some particular logical-based systems may directly use the developed techniques).

In this dissertation we extend the standard Boolean logic by formalising a number of logics that encode simultaneously different theories (of a circuit/system model), which represents an alternative to the classical approach of modelling each theory with Boolean variables and then relating the encoded theories with constraints over those variables. The logics developed typically have 2^n values, where n is the number of encoded theories, but other logics are also presented to explicitly consider unspecified values. The development of such logics culminates with the set algebra, thus generalising the encoding of an arbitrary number of theories in a single logic. All logics aim at modelling fault dependencies, as did the basic D-calculus of the ECAD area [Abramovici *et al.* 1990], already used in constraint programming [Simonis 1989]. We generalise not only the number of encoded theories, as explained, but also the number of faults a theory can have, thus, being no more restricted to single faults.

For each logic, we show how to model and solve different satisfaction and optimisation problems and we implement specialised constraint solvers, showing their applicability. All these implementations contributed to a workbench for a practical study of different consistency techniques. In addition, when implementing search strategies, a new technique, **iterative time-bounded search** (Chapter 3), was formalised and developed with significant results.

One such implemented constraint solver, ***Cardinal*** (Chapter 6), is a general set constraint solver which improves on existing solvers by adding *cardinality inferences*, which prove to be extremely useful by obtaining execution times orders of magnitude smaller on a number of problems. To such set solvers, we also show how attaching to set variables set functions other than cardinality, may turn general problems more declarative and efficient, by reasoning on such functions. We use an extended *Cardinal* for practical results over some benchmark problems.

In addition to the constructive approach of constraint programming we use a repairing approach in a tool, ***Maxx*** (Chapter 7), that we developed to solve the optimisation problem of maximising the number of unspecified bits in an input test vector. Since models with the developed logics and other approaches are not complete (in the sense that optimal tests can be unrecognised), we developed yet another logic to which we refer as **extended logic**. This logic takes into account the sources of unspecified values, in order to approach completeness in a practical way. Again, sets were used to denote dependencies on specified values in a **local search** method to improve solutions. Altogether, *Maxx* incorporates a number of multi-valued logics with branch-and-bound and local search. By obtaining better results than an efficient tool based on SAT and ILP, we show and discuss the usefulness of integrating branch-and-bound with local search.

In summary, the main contributions include:

- ❑ Development and formalisation of a number of multi-valued logics to model different problems, together with specialised CLP tools to solve them;
- ❑ Development and formalisation of a new efficient search technique: ITBS;
- ❑ Proposal of new benchmarks, namely for differential diagnosis;
- ❑ Generalisation of several logics into a single logic over sets, to encode an arbitrary number of theories;
- ❑ Formalisation of models for the different satisfaction and optimisation diagnostic-related problems;
- ❑ Generalisation of all problems to theories with multiple faults;
- ❑ Generalisation of all circuit problems to logical-based systems;
- ❑ Development and formalisation of a new efficient general set constraint solver, *Cardinal*, with especial inferences on sets cardinality;
- ❑ Presentation of extensions to *Cardinal*, generalising inferences over sets functions, with discussion and application over general combinatorial problems;
- ❑ Formalisation of an *extended logic*, more complete when handling unspecified values, by keeping track of their sources;
- ❑ Development of *Maxx*, a constraint tool incorporating different multi-valued logics

developed, to optimise test vectors and beat an existing efficient tool based on SAT and ILP;

❑ Integration of *local search* with *branch-and-bound* using another logic based on sets, with exemplification of the usefulness of such an approach on optimisation problems.

All in all, a number of research topics are open and we believe that the results and generalisations we obtain in several areas promise a wide research with many possible directions.

1.4.1 *Limitations*

While we show that the techniques developed in this dissertation generalise to logical-based systems, we restrict our examples to problems over digital combinational circuits with well-known benchmarks. The ECAD industry is very competitive and has many specialised and optimised tools devoted to solving each industry problem. It is already a very mature area; hence it is not our purpose to solve their specific problems or improve their solutions (although we present some competitive results, as recognised in the ECAD community, e.g. [Azevedo and Barahona 2000a, 2002]). Rather, we want to show the effectiveness of our approach, to these and other problems, by showing its potential and generalise it so that the already specialised techniques developed can be used in other domains. Also, we present new ideas and techniques that the industry may like to introduce in their systems in order to possibly improve them.

Therefore, we did not concentrate on heuristics, which are crucial for solving circuit problems. Also, during search, we relied on chronological backtracking of Prolog (usually the graphical Tcl/Tk version [Ousterhout 1994] of ECLiPSe over Windows). Some intelligent backtracking scheme such as dependency-directed backtracking [Stallman and Sussman 1977] would probably be much more efficient when trying to label a number of circuit input bits. Moreover, the underlying Prolog tool is not the best suited in terms of efficiency.

Problems and models presented throughout this dissertation assume that circuits are combinational. Nevertheless, solving problems over sequential circuits often involves some transformation into equivalent combinational circuits [Abramovici *et al.* 1990], which allows using the developed techniques for such circuits.

A limitation regarding fault modelling is that we only consider stuck-at faults. Although other types of permanent physical faults can be modelled with just stuck-at faults [Abramovici *et al.* 1990], circuits with intermittent and transient faults require extra techniques not dealt with in this dissertation. Nevertheless, the usual stuck-at faults are enough to represent any kind of fault in theoretical systems modelled by a set of propositional rules.

1.5 Overview

Since the practical examples of this dissertation are concentrated on circuit problems, we start by discussing circuit modelling in Chapter 2. A first problem is presented in Chapter 3, where we describe, model and solve the basic problem of generating tests (input vectors) for a circuit (particularly, for a specific fault) and discuss and compare our approach with competing alternatives. For this problem, we develop a CLP solver over a 4-valued logic (encoding fault dependencies) that uses a new search technique, namely, iterative time-bounded search. Such logic is then extended to an 8-valued logic that we formalise in Chapter 4 for the problem of differential diagnosis of two sets of faults.

In Chapter 5, we present different satisfaction and optimisation problems concerning multiple diagnoses (each a set of faults) showing how to model them with a logic over values formed by a pair of a set and a Boolean value (the particular problem of diagnosis is efficiently solved with such logic, as the experimental results document). We also show a transformation of such value pairs into single set values, that produce elegant models over a set algebra. To solve problems described by these and other models, we developed a general set constraint solver, *Cardinal*, formally described in Chapter 6, that is able to perform a number of especial inferences over sets cardinality. We present experimental results for general set problems and for the particular problem of differential diagnosis, where the superiority of *Cardinal* over existing solvers is especially evident. *Cardinal* extensions are also presented and discussed in this chapter by considering other set functions to solve other general problems, such as set covering.

Inspired in the reasoning over sets and over dependencies, in Chapter 7 we tackle another test optimisation problem by developing an extended logic and using local search (with sets of dependencies on specified values) together with a constructive approach to improve on an existing efficient tool to solve this problem. We incorporated a number of multi-valued logics to develop the new tool, *Maxx*, for which we present the improved results.

Finally, in Chapter 8 we show that all techniques generalise to consider logical-based systems (modelled by a set of propositional rules) instead of circuits. In addition to this important and desired result, we discuss the overall work covered in this dissertation together with possible future research, and present some final conclusions.

We organised this dissertation in a way that the multi-valued logics are presented in successive chapters in increasing order of the number of their values, as a consequence of encoding successively more theories. Often a described logic is a generalisation of a previous one; hence chapters should be read sequentially, although we try to keep each chapter with a reasonable amount of autonomy by having clear distinct objectives and referring related work described in more detail elsewhere in the dissertation (or outside it).

We add two appendices, basically with raw data, for ease of consultation and to obtain more detail on the most used sources of knowledge of this dissertation. Appendix A describes in more detail the circuit benchmarks used throughout the dissertation (an information made available on the internet via world wide web). Appendix B describes the multi-valued logics that were developed, with their purpose, meaning of values, and "truth" tables for the usual operations.

Circuit Modelling

To support all the stages of the life cycle of a digital system, it is convenient to have a model of it. Design, production and testing of a digital circuit largely depend on modelling, for it is the model that allows one to simulate the logical circuit with or without faults, to verify its correctness and generate tests for it. Hence, the way a circuit is *internally modelled* in a computer influences the application of algorithms for it.

This chapter addresses circuit- and fault-modelling approaches starting with a brief introduction in section 2.1, where we present general terminology and features of combinational circuits. In section 2.2 we discuss modelling of normal digital circuits for simulation of behaviour, and in section 2.3 possible circuit faults that may affect it are taken into account. Then, in section 2.4 we present a set of widely used circuit benchmarks and their characteristics, and in section 2.5 we describe our general modelling approach for such circuits and for the faults that may affect their behaviour.

2.1 Introduction

A combinational circuit C may be represented as a directed acyclic graph $<V_C, E_C>$ where nodes V_C are circuit gates together with primary inputs/outputs, and edges E_C connecting two nodes represent signal lines (also referred to as *nets*). This kind of representation allows the application of concepts and algorithms of graph theory [Harary 1969] and will be more thoroughly discussed in the next sections. Circuit gates are typically *and-*, *or-*, *xor-*, *nand-*, *nor-*gates and buffers. **Primary inputs** (PI) and **primary outputs** (PO) are the externally visible lines of the circuit. We can then define the (logic) **level** of an element node as the maximum path distance from PIs, and the level of a circuit as the maximum element level in it. Primary inputs level is 0.

When a circuit has n ($n > 1$) different edges sharing one source node, it means that a net propagates a signal from one source to n destinations. Such a signal is said to have **fanout** and may be represented by a fanout node like the one in Figure 2.1, where the *stem* is an edge from the source node and the *n fanout branches* are the edges going to the corresponding destinations (or loads). A **fanout-free** circuit has no signals with fanout (or, equivalently, every signal has a fanout count of 1), being thus represented as a tree (a particular case of graph) for each PO.

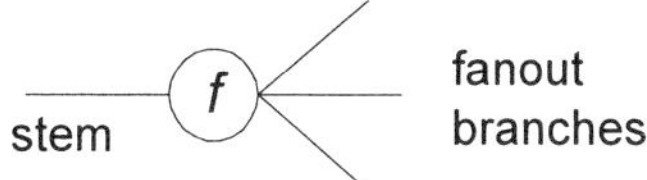

Figure 2.1. Fanout node

When different paths from the same signal reconverge at the same component (i.e. there are two different paths between two graph nodes) we have **reconvergent fanout**, which complicates the tasks of test generation and diagnosis, as we will see later.

Conversely, the **fanin** nodes of a node x are the source nodes of edges sharing x as destination.

In addition, for a circuit node x in V_C, we define its **transitive fanout** as the set of all nodes y such that there is a path connecting x to y. Similarly, the **transitive fanin** of x is the set of all nodes y such that there is a path connecting y to x.

For a path with no *xor*-gates, we can also define its **inversion parity** as the number of its inverting gates (*not-*, *nand-* and *nor*-gates), modulo 2. This will be useful for some algorithms.

If the circuit contains *xor*-gates and one wants to make use of such algorithms then these gates should be modelled as a sub-circuit of other basic gates (where different paths may occur). The reason is that an *xor*-gate with inputs a and b may be seen as a conditional inverter (where if, say, a takes value 1 then b is inverted, otherwise the gate outputs b, as a simple buffer).

2.2 Logic Simulation

The behaviour of a system is defined by the I/O mapping performed by its *black box* model where the information carried by its inputs is processed to produce some output. Such value transformation, separated from the time domain where it occurs, is referred to as **logic function**, being its representation a **functional model** of the system.

A combinational circuit with n binary inputs outputting function $Z(x_1,x_2, ..., x_n)$ can be functionally modelled at the logic level by a **truth table** with 2^n entries. The compact representation may then be essential to save memory, but not necessarily since in the full 2^n array A only the output values have to be stored. The storage of input combinations may be suppressed by implicitly considering array entries in increasing order of binary words formed by the concatenation of the input values. For instance, $A(0)=Z(0,0, ...,0)$, $A(1)=Z(0,0, ...,1)$ and $A(2^n-1)=Z(1,1, ...,1)$. For a circuit with m outputs, an entry in the array is an m-bit vector defining the output values. In practice, functional models are only used for small circuits because, on the one hand, they are impractical for large circuits and, on the other hand, they do not represent the internal components whose states may be the subject of interest (e.g. whether they are faulty).

A **structural model** describes a circuit as a collection of interconnected components or **primitive elements** of various **types**. Such model can be represented graphically as a *schematic diagram* where component types are typically AND-, OR- and other gates with special shapes to symbolise them (Figure 2.2, inputs on the left, output on the right).

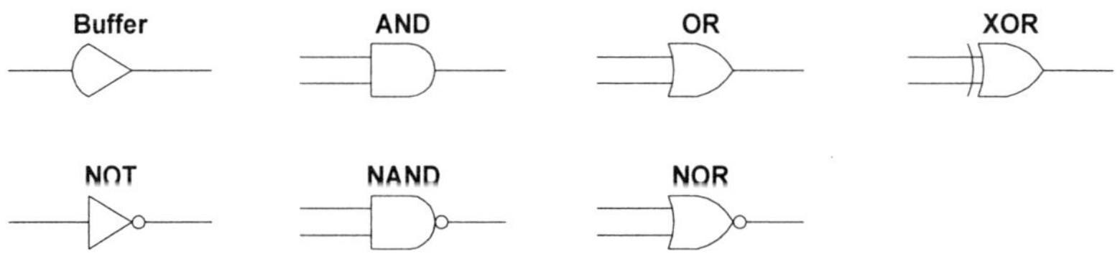

Figure 2.2. Basic gates

Gates are components which are small enough to be usefully represented by functional models, corresponding to some **Boolean logic** operation, usually in the form of truth tables with one input (NOT-gate) or two inputs (AND-gate and OR-gate) as exemplified in Figure 2.3. Thus, in practice, structural and functional modelling are generally intermixed.

NOT			AND	0	1		OR	0	1
0	1		**0**	0	0		**0**	0	1
1	0		**1**	0	1		**1**	1	1

Figure 2.3. Boolean logic truth tables

The logic function of, say, a binary *and*-gate can then be represented by its truth table as in Figure 2.4 (a) or, more compactly, in Figure 2.4 (b), where x stands for an unspecified or "don't care" value.

<table>
<tr><td colspan="2"> </td><td> </td><td colspan="2"> </td><td> </td></tr>
</table>

A	**B**	**Z=A.B**		**A**	**B**	**Z=A.B**
0	0	**0**		0	x	**0**
0	1	**0**		x	0	**0**
1	0	**0**		1	1	**1**
1	1	**1**				

(a)   (b)

Figure 2.4. And-gate truth table

The function of a circuit can also be modelled by a graph called a **binary decision diagram** [Lee 1959, Akers 1978]. The value of the output is determined by sequentially examining input values on traversal of the graph. Figure 2.5 (a) shows the complete binary tree for the *and*-gate. Starting at the top node, we take the left or the right branch according to the value (0 or 1, respectively) of the corresponding input. The value at the exit branch is the value of the output Z for those inputs. These diagrams can be simplified to yield smaller graphs as in Figure 2.5 (b) for the same gate, the result in this case being still a simple binary tree. Now, a variable at an exit branch means that the function has the value of that input variable (e.g. for $Z=A.B$, if $A=1$ then $Z=B$).

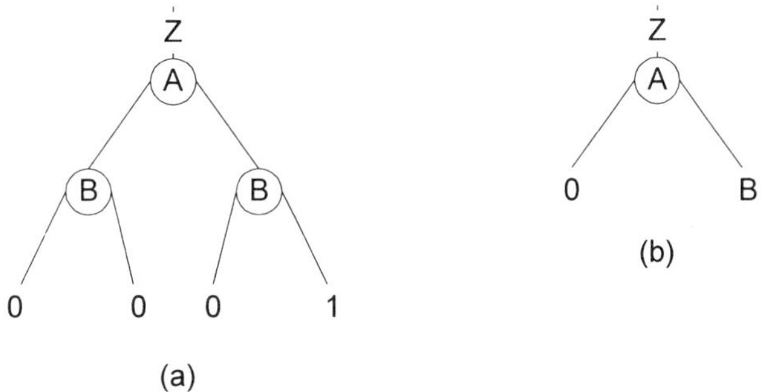

Figure 2.5. Binary decision diagrams for the and-gate

So far we have only seen models as data structures that may be interpreted by model-independent programs. Nevertheless, it is also possible to model the function of a circuit directly by a program in **code-based modelling**, which is the approach used for the primitive elements in a structural model. It is generally easier to develop models based on data structures, but code-based models may be more efficient since there is no need to interpret a data structure.

For example, the circuit of Figure 2.6 can be modelled in the C programming language, considering only binary values, as follows:

$$E = A \,\&\, B \,\&\, C$$
$$F = \sim D$$
$$Z = E \mid F$$

Such type of model may be also referred to as *compiled-code* model, since it is compiled into machine code.

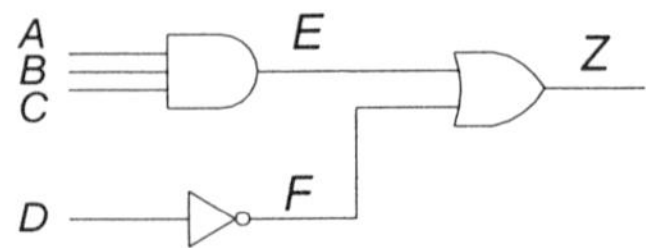

Figure 2.6. Example circuit to model as code

An application may also automatically generate a functional model from a structural model of a circuit. For the same example of Figure 2.6, one can get the following resulting model in assembly code:

```
LDA  A        /* load accumulator with value of A */
AND  B        /* compute A.B */
AND  C        /* compute A.B.C */
STA  E        /* store partial result */
LDA  D        /* load accumulator with value of D */
INV           /* compute D̄ */
OR   E        /* compute A.B.C + D̄ */
STA  Z        /* store result */
```

Such a model applies the standard Boolean functions and requires specified Boolean values. In practice, however, for many applications, some values may be unknown or unimportant (i.e. "don't know" / "don't care" values). These **unspecified** values are part of the problem and must be considered as logic values themselves, thus extending the Boolean logic to a **3-valued logic** (Figure 2.7) that the model has to take into account. For instance, if the input of a *not*-gate is unspecified (unknown) so is its output. Similarly, if an *and*-gate's input takes value 0, so does the output, and we "don't care" about the other input, which may remain unspecified.

NOT	
0	1
1	0
x	x

AND	**0**	**1**	**x**
0	0	0	0
1	0	1	x
x	0	x	x

OR	**0**	**1**	**x**
0	0	1	x
1	1	1	1
x	x	1	x

Figure 2.7. Boolean logic extended with unspecified values

Non-Boolean logics make code-based models more problematic due to its increased complexity and the lack of built-in operations over such logics in digital computers.

In addition, the circuit may be faulty and have an altered logic function. Consequently, for tasks other than circuit design, the model must also consider the presence of faults, which is the topic of the next section.

2.3 Fault Modelling

Different physical faults in a circuit may be modelled by logical faults that represent their effect. Hence, we can handle fault analysis logically, and independently from technology, by the use of a **logical fault model**. Also, tests derived for logical faults (test generation will be discussed in the next chapter) may be used for physical faults whose effect on circuit behaviour is not completely understood or is too complex to be analysed [Hayes 1977].

For off-line testing, we will only deal with **permanent faults** since modelling intermittent and transient faults requires statistical data on their probability of occurrence, which are usually not available. Given such a permanent fault and a model of the circuit, it is possible to determine its logic function in the presence of the fault. Thus we also define faults as **structural faults** or

functional faults according to the circuit model used. Structural faults only modify the value of interconnections among components assumed to be fault-free, whereas functional faults change the truth table of a component. The typical faults affecting interconnections are their breaking, known as **opens**, and unwanted connections of points, known as **shorts**.

A short between a signal line and ground or power can make the signal remain at a fixed voltage level. Such fault is logically modelled as the signal being **stuck at** the corresponding fixed logic value v ($v \in \{0,1\}$), and it is denoted by *s-a-v*, i.e. the line has always the same logic value v, regardless of the inputs that would normally affect it.

An open may also appear as a stuck fault since it usually makes the input that has become disconnected to assume a constant logic value (Figure 2.8). This is also the result of a physical fault internal to the component driving the line. In edge-pin testing, the two cases may be treated as the entire signal line being stuck. Hence, the single logical fault of a line *s-a-v* ($v \in \{0,1\}$) represents many different physical faults. The restriction of considering only faults associated with the I/O pins of components is referred to as *pin-fault model*.

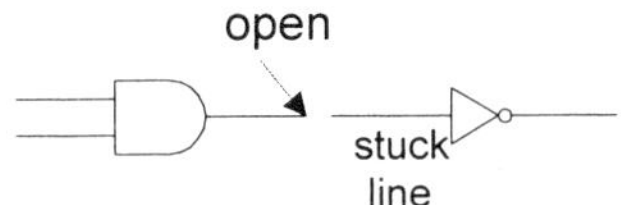

Figure 2.8. Stuck fault caused by an open

In this way, faults may be incorporated in a logic model by extra truth tables for faulty gates, as in the simple Table 2.1 for an *and*-gate s-a-0.

Table 2.1. And-gate s-a-0 truth table

AND s-a-0	0	1	x
0	0	0	0
1	0	0	0
x	0	0	0

The standard fault model is then the classical single-stuck fault (**SSF**) model, the first and most widely studied and used. The SSF model adopts the **single-fault assumption** that there is at most one logical fault in the system. This is due to the *frequent testing strategy* in which the system is tested often enough to keep a negligible probability of more than one fault developing between two consecutive testing experiments. However, frequent testing may be insufficient to avoid the occurrence of multiple faults, since physical faults sometimes correspond to multiple logical faults. Also, in newly manufactured systems prior to their first testing, multiple faults are likely to exist. An undetected single fault in a testing experiment may also lead to a multiple fault if a second single fault occurs between two testing experiments. Nevertheless, tests designed for single faults can detect many multiple faults as well as many non-classical faults.

The SSF model can thus be applied to any structural model, independently of technology, and represents many different physical faults [Timoc *et al.* 1983]. Another advantage is that the total number of SSFs in a circuit is small compared to other fault models, and not all of the faults have to be explicitly analysed in edge-pin testing. Moreover, SSFs can be used to model other types of faults at the cost of increasing the size of the circuit model.

2.4 Benchmarks

Benchmark circuits provide a common workbench for the scientific community to test and compare different applications for a variety of ECAD problems. Typically, a benchmark structural model is expressed in some *connectivity language* specifying the I/O lines and signals of the system and its components. This is the case with the ISCAS'85 benchmark suite [ISCAS 1985, Brglez and Fujiwara 1985], the well known and most used set of benchmark circuits in the area. It comprehends a total of 11 combinational circuits (*c17*, *c432*, *c499*, *c880*, *c1355*, *c1908*, *c2670*, *c3540*, *c5315*, *c6288* and *c7552*), where the number in a circuit name indicates the number of signal lines (nets) in the circuit. The simplest of them, *c17*, is described below as an example, and corresponds to the schematic diagram of Figure 2.9:

```
INPUT(1gat)
INPUT(2gat)
INPUT(3gat)
INPUT(6gat)
INPUT(7gat)

OUTPUT(22gat)
OUTPUT(23gat)

10gat = nand(1gat,   3gat)
11gat = nand(3gat,   6gat)
16gat = nand(2gat,   11gat)
19gat = nand(11gat,  7gat)
22gat = nand(10gat,  16gat)
23gat = nand(16gat,  19gat)
```

Primary inputs and primary outputs are explicitly stated. The definition of gates follows the form *output = type(input_list)* which specifies the gate type (whose functional model is assumed to be known) and its I/O terminals (a single output and functionally equivalent inputs). Signal names in these terminals implicitly describe interconnections. For instance, primary input *3gat* is connected to *nand*-gates *10gat* and *11gat*.

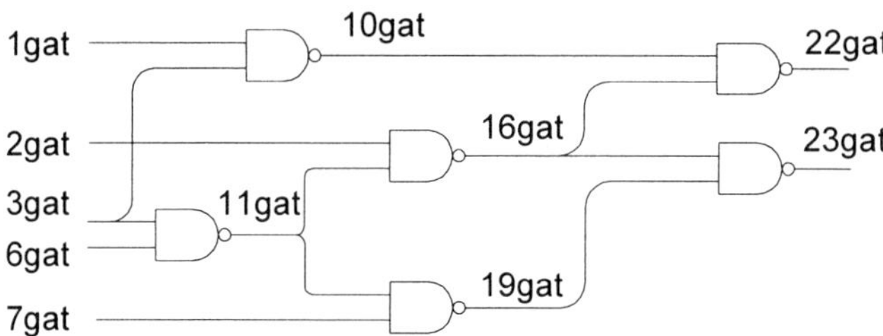

Figure 2.9. *C17* ISCAS'85 example circuit

Applying the previous definitions, we check that the level of *c17* (graph in Figure 2.10) is thus 3, corresponding to the longest path (e.g. *3gat* → *11gat* → *16gat* → *22gat*).

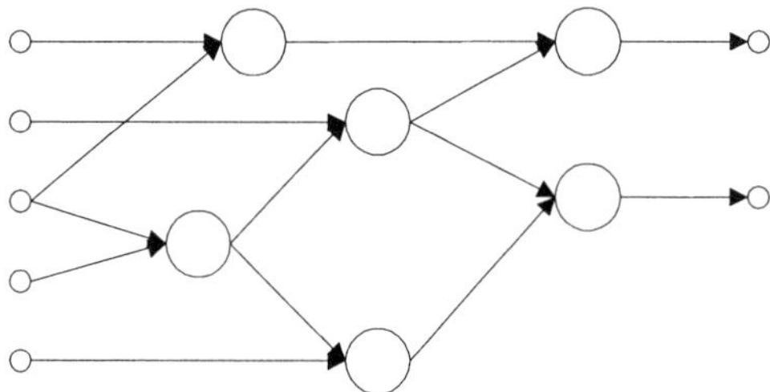

Figure 2.10. Graph representation of *c17*

It is easy to verify that *c17* presents reconvergent fanout, with (for instance) two paths from *11gat*

(*11gat* → *16gat* → *23gat* and *11gat* → *19gat* → *23gat*) reconverging at node *23gat*.

Remembering the definitions of transitive fanin and fanout we see that, for example, the transitive fanout of *11gat* is {*16gat, 19gat, 22gat, 23gat*} and the transitive fanin of *22gat* is {*1gat, 2gat, 3gat, 6gat, 10gat, 11gat, 16gat*}.

Table 2.2 shows some statistics concerning each benchmark, including the number of PIs, POs, gates, circuit level, and average and maximum fanin and fanout, where a gate fanin is the number of its inputs.

Table 2.2. Lines' statistics of ISCAS benchmarks

circuit	PI	PO	gates	level	avg fanin	max fanin	fanout stems	fanout lines	avg fanout	max fanout
c17	5	2	6	3	2.00	2	3	6	1.27	2
c432	36	7	160	17	2.10	9	89	236	1.75	9
c499	41	32	202	11	2.02	5	59	256	1.81	12
c880	60	26	383	24	1.90	4	125	437	1.70	8
c1355	41	32	546	24	1.95	5	259	768	1.87	12
c1908	33	25	880	40	1.70	8	385	995	1.67	16
c2670	233	140	1193	32	1.74	5	454	1244	1.55	11
c3540	50	22	1669	47	1.76	8	579	1821	1.72	16
c5315	178	123	2307	49	1.90	9	806	2830	1.81	15
c6288	32	32	2416	124	1.99	2	1456	3840	1.97	16
c7552	207	108	3512	43	1.75	5	1300	3833	1.68	15

This set of circuits is fairly heterogeneous which, as is also visible by the gates' distributions shown in Table 2.3, makes them suitable for comparing different approaches for all kinds of problems. It is noteworthy that *c1355* is just an expansion of *c499*, in replacing each *xor*-gate by 4 *nand*-gates, with a few buffers added.

Table 2.3. Gates' statistics of ISCAS benchmarks

Circuit	buffer	not	and	nand	or	nor	xor	Total
c17				6				6
c432		40	4	79		19	18	160
c499		40	56		2		104	202
c880	26	63	117	87	29	61		383
c1355	32	40	56	416	2			546
c1908	162	277	63	377		1		880
c2670	196	321	333	254	77	12		1193
c3540	223	490	498	298	92	68		1669
c5315	313	581	718	454	214	27		2307
c6288		32	256			2128		2416
c7552	534	876	776	1028	244	54		3512

For test generation problems, the ISCAS benchmarks include a set of single stuck-at faults (SSF) for each circuit (Table 2.4). Benchmark faults are only a subset of all the possible faults in a circuit (which are twice the number of nets, since any net may be stuck-at-0 or stuck-at-1), in fact, they constitute a **collapsed fault set** [Brglez and Fujiwara 1985]. The complete set of possible SSFs is collapsed to a smaller set since faults that are functionally equivalent to some other may be discarded. Two faults *f* and *g* are said to be **functionally equivalent** if the circuit presents always the same behaviour under the presence of fault *f* or *g*. For instance, an *and*-gate s-a-0 is functionally equivalent to any input *i* s-a-0 (as long as *i* has no fanout). In addition, the fault set may be further collapsed by considering the **fault dominance** relation [Abramovici *et al.* 1990].

Fault f dominates g iff any test that detects* g also detects f (on the same primary outputs), i.e. the set of tests that detect f contains g's. As an example, for $z = x$ *and* y, the single test that detects $g = x$ s-a-1 is $t = 01$ that also detects $f = z$ s-a-1. Fault f thus dominates g and can be discarded.

Table 2.4. ISCAS circuits: Fault sets

	c17	c432	c499	c880	c1355	c1908	c2670	c3540	c5315	c6288	c7552
Faults	22	524	758	942	1574	1878	2746	3425	5350	7744	7550

2.5 Our Modelling Approach

We have adopted a structural gate-level code-based modelling for the ISCAS benchmarks. It is a *gate-level* model since the lowest-level primitive components are gates (including the *xor*-gate) that cannot be structurally decomposed. The logic value of each signal line is given by a different variable. Hence, to allow for different values in different fanout branches from the same stem (due to circuit faults), each branch is replaced by a buffer outputting a different variable. The same happens with PIs as Figure 2.11 illustrates with the *c17* circuit. It is a code-based model because we have used a programming language (PROLOG [Lloyd 1988, Sterling and Shapiro 1994]), as well as its constraint logic programming extensions to more efficiently tackle different circuit problems.

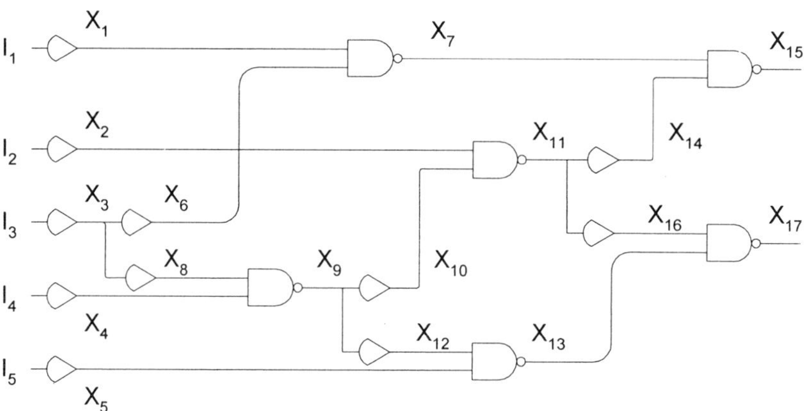

Figure 2.11. *C17* circuit with added buffers

This circuit is then basically modelled by predicate *circuit / 4* (4 arguments) below (Figure 2.12) as a sequence of 17 gates outputting 17 variables ($X1$ to $X17$). These variables are uniquely identified by an internal name given by a number i ($i \in \{1..17\}$). Each gate is of the type shown in the 1st argument of predicate *gate / 6*, has a list of inputs (2nd argument) and a single output (3rd). Such gate appears in the order of its output number i (4th argument) in a way that for any other gate (output number) j, if $i < j$ then i does not depend on j, i.e. being $l(i)$ the level of the gate, if there is a possible path connecting i and j then $l(i) \leq l(j)$. Hence, either there is no path between i and j or the gate with the lowest level appears first in the circuit model predicate. We thus guarantee that, when reaching some gate, its transitive fanin has already been handled.

---

* Fault detection and test generation will be discussed in more detail in the next chapter

```
%------------------------------------------------------
%circuit(?InputList, ?OutputList, +InfoIn,-InfoOut)
%--
circuit([I1,I2,I3,I4,I5], [X15,X17], A0,A17):-
    gate(inpt, [I1],      X1,   1,  A0, A1),
    gate(inpt, [I2],      X2,   2,  A1, A2),
    gate(inpt, [I3],      X3,   3,  A2, A3),
    gate(inpt, [I4],      X4,   4,  A3, A4),
    gate(inpt, [I5],      X5,   5,  A4, A5),
    gate(buff, [X3],      X6,   6,  A5, A6),
    gate(nand, [X1,X6],   X7,   7,  A6, A7),
    gate(buff, [X3],      X8,   8,  A7, A8),
    gate(nand, [X8,X4],   X9,   9,  A8, A9),
    gate(buff, [X9],      X10, 10,  A9, A10),
    gate(nand, [X2,X10],  X11, 11,  A10,A11),
    gate(buff, [X9],      X12, 12,  A11,A12),
    gate(nand, [X12,X5],  X13, 13,  A12,A13),
    gate(buff, [X11],     X14, 14,  A13,A14),
    gate(nand, [X7,X14],  X15, 15,  A14,A15),
    gate(buff, [X11],     X16, 16,  A15,A16),
    gate(nand, [X16,X13], X17, 17,  A16,A17).
```

Figure 2.12. circuit/4 predicate for *c17*

The circuit predicate receives the lists of PIs and POs, and, through the last two arguments, processes some application-dependent information that is passed along the gates (predicates), such as a fault or a list of possible faults. It is up to a specific implementation of the *gate / 6* predicate to handle this information according to a goal. Different *gate / 6* implementations are shown in Figure 2.13, Figure 2.14, Figure 2.15 and Figure 2.16 to illustrate some of the possible applications.

The code in Figure 2.13 may be used for simulation of the circuit's logic function. Here, the predicate consists of a simple Boolean logic table lookup. The 2 simple alternative clauses represent opposing alternatives, and, if inputs are still variables, choice-points are created relying on the PROLOG backtracking mechanism to eventually reach a possible solution on some possibly given circuit outputs.

```
%------------------------------------------------------
%gate(+Type, ?InputList, +Output, +Name, +InfoIn,-InfoOut)
% Boolean simulation
%--
gate(not, [0], 1, _, X,X).
gate(not, [1], 0, _, X,X).
```

Figure 2.13. Gate predicate for Boolean simulation

Figure 2.14 represents the extension to a different specific logic where, if predicate *multi_valued_logic_not / 2* posts any constraint on the gate signals *In* and *Out* then the gate predicate may also be used as some specialised constraint over that logic.

```
%------------------------------------------------------
%gate(+Type, ?InputList, +Output, +Name, +InfoIn,-InfoOut)
% Multi-valued logic simulation
%--
gate(not, [In], Out, _, X,X):- multi_valued_logic_not(In, Out).
```

Figure 2.14. Gate in a multi-valued logic

For diagnostic goals, the code of Figure 2.15 may be used for simulation of a faulty circuit. The gate predicate checks whether the gate is to be treated as faulty (SSFs are given and processed in a sorted list using the last 2 arguments).

```
%------------------------------------------------------------
%gate(+Type, ?InputList, +Output, +Name, +FaultsIn,-FaultsOut)
% Faulty circuit simulation
%--
gate(not, [_], 0, Name, [Name:s-a-0|Fs],Fs):- !.
gate(not, [_], 1, Name, [Name:s-a-1|Fs],Fs):- !.
gate(not, [0], 1, _,    Fs,Fs).
gate(not, [1], 0, _,    Fs,Fs).
```

Figure 2.15. Possible faulty gate

Test generation for some fault(s) may be performed using constraints as in Figure 2.16. Here, the gate predicate is used as a specialised constraint over a specific logic (by means of an arithmetic constraint).

```
%------------------------------------------------------------
%gate(+Type, ?InputList, ?Output, +Name, +InfoIn,-InfoOut)
% Constraint
%--
gate(not, [In], Out, _, X,X):- Out #= 1-In.
```

Figure 2.16. Gate as a constraint

As an example, using an implementation such as that of Figure 2.15 to model stuck gates, and applying it to *c17* with *nand*-gate 7 s-a-0 (Figure 2.17), we can evaluate the goal *circuit([0,0,0,0,0], Out, [7:s-a-0], [])* to obtain *Out=[1,0]*. Notice that evaluating the circuit with no faults, *circuit([0,0,0,0,0], Out, [], [])* would yield *Out=[0,0]*.

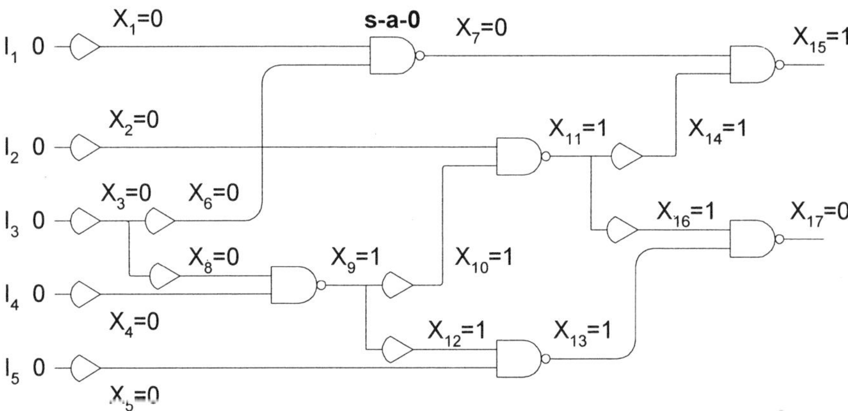

Figure 2.17. *C17* circuit with a stuck gate

The model is thus flexible enough to be used by many different types of applications, and is not restricted to Boolean logic. The introduction of extra buffers does not jeopardise efficiency since they will only have special treatment if potentially faulty, otherwise a simple variable unification is performed between input and output.

If necessary, the benchmark's faults are available as a list of *i/b* pairs, where *i* is the internal number of the gate and *b* is the Boolean stuck value. For example, *c17* fault set which is originally given as the 22 faulty signals of the left column of Table 2.5 is translated into the set of 22 SSFs in the form of stuck gates on the right column.

Table 2.5. Translation of *c17* fault set into stuck gates

Faulty signals	Stuck gates
1gat/1, 2gat/1, 3gat /0 /1,	1/1, 2/1, 3/0, 3/1,
6gat/1, 7gat/1, 10gat/1,	4/1, 5/1, 7/1,
3gat->10gat /1, 11gat /0 /1,	6/1, 9/0, 9/1,
3gat->11gat /1, 16gat /0 /1,	8/1, 11/0, 11/1,
11gat->16gat /1, 19gat/1,	10/1, 13/1,
11gat->19gat /1, 22gat /0 /1,	12/1, 15/0, 15/1,
16gat->22gat /1, 23gat /0 /1,	14/1, 17/0, 17/1,
16gat->23gat /1	16/1

In addition, to more efficiently solve the different possible applications, the graph data structure of the circuit is also available in our model conveying the information of Table 2.6 which includes the gates' external names for more user-friendly interfaces.

Table 2.6. *C17* graph information

Number	Name	Type	Inputs	Fanouts
1	1gat			7
2	2gat			11
3	3gat			6, 8
4	6gat			9
5	7gat			13
6	3gat_10gat	buffer	3	7
7	10gat	nand	1, 6	15
8	3gat_11gat	buffer	3	9
9	11gat	nand	4, 8	10, 12
10	11gat_16gat	buffer	9	11
11	16gat	nand	2, 10	14, 16
12	11gat_19gat	buffer	9	13
13	19gat	nand	5, 12	17
14	16gat_22gat	buffer	11	15
15	22gat	nand	7, 14	
16	16gat_23gat	buffer	11	17
17	23gat	nand	13, 16	

There is a functional model for each gate type modelled by a logic table which depends on the logic used. Of course, faulty gates do not obey to such "normal" tables and, in principle, extra logic tables should be considered for each type of faulty gate (given by the gate type and the stuck value). One could think that it would be enough to set the gate's output to the stuck value, independently of the gate type, thus avoiding new truth tables. This is certainly true to the extent that we are sure that the gate is really stuck and that we are working with simple Boolean logic. In general, however, the faults to handle are only potential faults that we want to confirm or infirm, so the output function is not certain yet. In many cases, the logic is extended to extra values that carry additional information on the possible physical Boolean variables. Then, when the inputs and output of a potential faulty gate are not known, the logic function is seen as a relation that may "pass" information forward or backward. Logic tables for these gates are thus essential.

To simplify, we only consider stuck buffers. A general stuck gate may be seen just as a normal gate followed by a fictitious stuck buffer. Then, for each target fault (a stuck gate) we add such a buffer, which we refer to as an **S-buffer** (Figure 2.18) [Azevedo and Barahona 1998]. In this way, all gates are normal and are thus modelled by the logic table (possibly expressed by a constraint) corresponding to its type (*and, or, ...*). Only S-buffers may be stuck, hence they are modelled by a different relation.

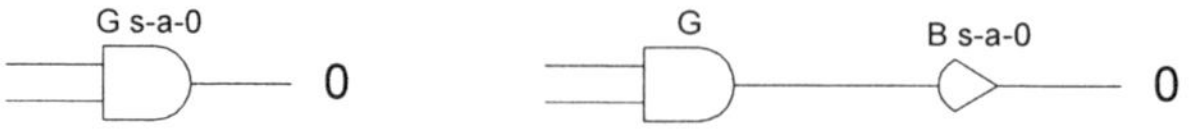

G s-a-0   0   G   B s-a-0   0

Figure 2.18. Introduction of an S-buffer B s-a-0

When evaluating a *gate / 6* predicate, the logic table corresponding to its type is treated normally and, with the information passed through all the gates by the two final *gate* arguments, it is checked whether it is a potential faulty gate, in which case the logic table of the S-buffer is processed.

The use of S-buffers, their logic tables and the extended logics will be discussed and made clearer in the next chapters.

2.6 Summary

This chapter presented combinational digital circuits as the global subject of the problems and examples that will be covered throughout this thesis, and our general modelling approach for them considering the possible faults that may affect their behaviour.

The next chapter presents a basic problem (testing) involving some possible fault(s) in the circuit, and discusses and compares approaches and algorithms for circuit or fault testing.

Test Patterns

Since a circuit may present an unintended behaviour (i.e. output an incorrect function), it is important to know whether it is normal or faulty, so that its output can be trusted. A correct output (as predicted by its fault-free model) for a given input is, however, generally insufficient to conclude that the circuit is normal, since typically faults only affect a fraction of the circuit function. Thus, only some input-output configurations highlight the presence of specific fault(s). Since it is impractical to test a circuit for all possible inputs, dedicated or limited tests are usually performed to test stuck lines (which may however skip detection of some possible faults).

Choosing such tests is the subject of this chapter, where we start introducing the notion of test patterns in the next section, and discuss its generation in section 3.2. In section 3.3 we present some modelling approaches and algorithms, including the introduction of multi-valued logics to model fault dependencies in a circuit. Constraint reasoning, and its inclusion in a solver we developed for test generation, is discussed in section 3.4. Heuristics to guide problem solving are discussed in section 3.5 before our alternative, iterative time-bounded search, is presented together with results and conclusions in section 3.6.

3.1 What are Test Patterns ?

To check whether a system behaves as expected (i.e. whether its response function is the same of its model or its design specifications), we need to test it with certain stimuli and verify its response. For a digital circuit, these stimuli are its input bits with possible logic values 0 or 1, thus forming an input vector referred to as a *test pattern*. The number of test patterns of a circuit is exponential on the number of primary inputs since each combination of input values is a potential test pattern. Therefore, only a restricted number of test patterns (also called a *test set*) is usually considered for the circuit under test (CUT), since it is expensive and even impractical to consider all patterns. In addition, most of the times a small test set is enough to test all possible detectable single stuck faults (SSFs). For example, for the circuit of Figure 3.1, the test set $T =$ {000, 001, 111} is a *complete detection test set* for all stuck gates since any of the 8 possible SSFs (4 gates times 2 Boolean values) can be detected by applying at least one test pattern $t \in T$.

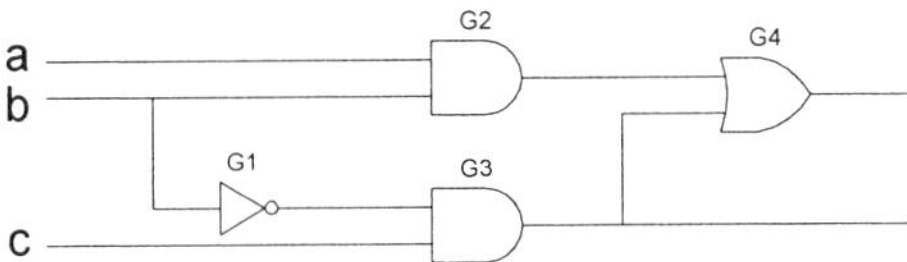

Figure 3.1. Example CUT

For instance, test $t=001^{*}$ detects SSF $f = G1$ *s-a-0*, as shown in Figure 3.2, since the circuit response function changes under f.

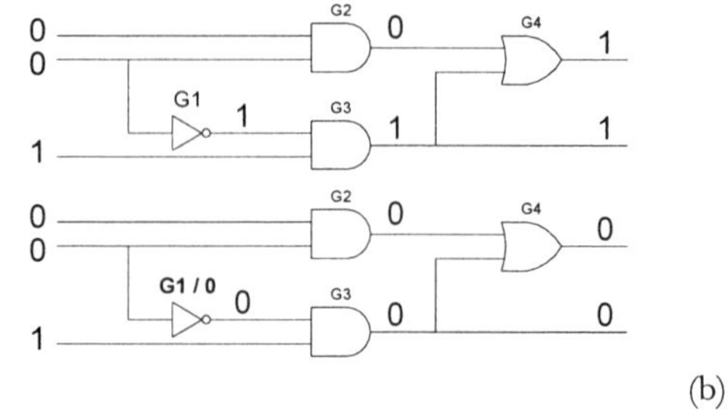

(a)   (b)

Figure 3.2. Expected circuit behaviour (a); and circuit response with *G1* s-a-0 (b)

Table 3.1 below shows possible tests *t* from set *T*, for each stuck gate.

Table 3.1. Tests for possible stuck gates

SSF	t
G1/0	001
G1/1	111
G2/0	111
G2/1	000
G3/0	001
G3/1	000, 111
G4/0	001, 111
G4/1	000

A variety of techniques exist to generate all elements (i.e. test patterns) of a test set for a given circuit. The techniques depend on the goals to be achieved as discussed in the following sections.

3.2 Test Generation

When the goal is to detect specific faults in the circuit, a test with appropriate inputs must be found so that the effects of the faults are apparent on the circuit output. This *Test Generation* (TG) is said to be fault-oriented. TG is an NP-complete problem [Ibarra and Sahni 1975], thus a number of algorithms have been developed to tackle it, due to its importance in industry. For the process of finding a test set for the SSFs in a circuit, generally there is an *Automatic Test Generation* (ATG) system that uses such an algorithm and a model of the circuit. The generated test set is also useful to detect many physical and design errors [Abadir *et al.* 1988].

For economic and quality reasons, the test set should, as much as possible, be: *a)* cheap (i.e. generated in a small amount of time); *b)* short; and *c)* have a high fault coverage, where the fault coverage for detectable faults (some faults may have no possible detection) is computed by

$$\frac{\textit{number of detected faults}}{\textit{total number of faults - number of undetectable faults}}$$

(i.e. it should detect as much faults as possible — ideally it should be a complete detection test set.)

In practice, it is not possible to achieve all these optimisation goals, hence user-specified limits are imposed on them. Of course, some goals may be more important than others since the fault coverage directly affects the quality of the product, which may be the key factor. Also, the generation time is not as costly as a long test set since longer test sets imply longer testing times, which may be repeated often, as opposed to the former, which is a one-time expense.

Typically, an ATG system in a first phase generates tests in a pseudo-random way providing a low-cost initial set that may normally detect from 50 to 80 percent of the faults. This first phase of ATG is said to be **fault-independent** since tests are not generated for specific faults. In a subsequent second phase, a **fault-oriented** test generation algorithm is applied for the undetected faults [Breuer 1971, Agrawal and Agrawal 1972]. In this chapter we only focus on this more difficult fault-oriented test generation.

Test vectors produced by a fault-oriented TG algorithm are, in general, partially specified, i.e. some input bits may have an unspecified or "don't care" value denoted by an x. For example, an ATG system may generate tests $0x1$ and $x11$. Unspecified values are important during TG when modelled circuit signals have not yet been assigned a value, and also as final generated PI values since they may allow a subsequent compaction of a test set. For instance, the example tests $0x1$ and $x11$ could be combined and replaced by the single test 011.

3.3 TG Modelling Approaches and Algorithms

Using the notation and definitions in [Abramovici *et al.* 1990], let Z denote the logic function of a combinational circuit N. We will denote by t a specific input vector, and by $Z(t)$ the response of N to t. For a multiple output circuit $Z(t)$ is also a vector. The presence of a fault f transforms N into a new circuit N_f. Here we assume that N_f is a combinational circuit with function $Z_f(x)$. The circuit is tested by applying a sequence T of test vectors t_1, t_2, ..., t_m, and by comparing the obtained output response with the (expected) output response of N, $Z(t_1)$, $Z(t_2)$, ..., $Z(t_m)$.

Definition 3.1: A test (vector) t **detects** a fault f iff $Z_f(t) \neq Z(t)$.

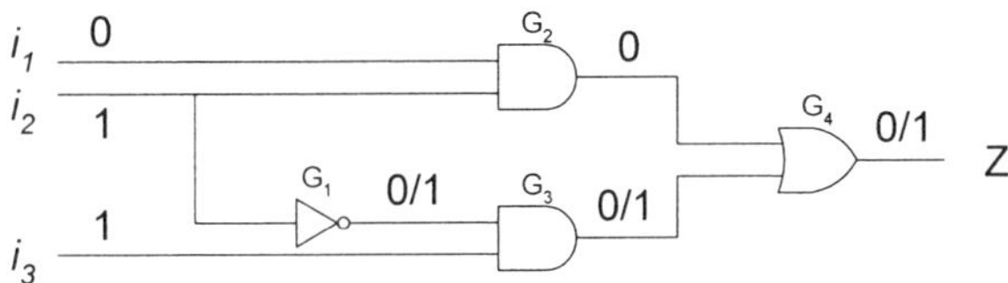

Figure 3.3. Test for SSF G_1 s-a-1

In the example of Figure 3.3, test $t = 011$ detects fault $f = G_1$ s-a-1. Signals that have different values without and with f present in the circuit, are shown in the form of a *composite logic value* v/v_f, where v is the signal value in the fault-free circuit N and v_f is the value in the faulty circuit N_f defined by the target fault f. The fault is detected since the output values in the two cases are different.

A fault f is said to be **detectable** if there exists a test t that detects f, otherwise it is *undetectable*. It is possible, in principle, to generate a test for a fault (or prove that it is undetectable) by trying successively different input vectors and simulating the good and faulty circuits until their responses differ in at least one output bit. If all possible input vectors are unsuccessful, the fault is undetectable. This *generate-and-test* approach is of course impractical for circuits with many input bits, since its time complexity is exponential on the number of such bits.

3.3.1 Algebraic Models / Algorithms

In the circuit of Figure 3.3, the general output function is given by the single output $Z = \overline{i_1.i_2} + i_2.i_3$. A test t that detects a fault f makes $Z(t)=0$ and $Z_f(t)=1$ or vice versa, i.e. $Z(t) \oplus Z_f(t) = 1$. For fault $f = G_1$ s-a-1, $Z_f = \overline{i_1.i_2} + i_3$ and $Z \oplus Z_f = 1$ reduces to $\overline{i_1.i_2}.i_3 = 1$, which enforces $i_1=0$, $i_2=1$, $i_3=1$, corresponding to the single test $t = 011$. Such method of representing the circuit by Boolean equations is called an *algebraic* method [Breuer and Friedman 1976].

Propositional Satisfiability (SAT) solvers [Larrabee 1992, Silva and Sakallah 1997] can be used for algebraic methods but these are considered impractical for large circuits [Abramovici *et al.* 1990].

3.3.2 Topological Methods

The usual TG methods can be characterised as *topological* as they are based on a structural model of a circuit.

Figure 3.3 illustrates two basic concepts used in this approach for fault detection:

- a test t that detects a fault f **activates** f, i.e., generates an error (a fault effect) by creating different v and v_f values at the site of the fault (a 0-value in the normal circuit and a 1-value in the faulty circuit or vice-versa, as in the output of faulty G_1 of Figure 3.3).

- t **propagates** the error to a primary output w, that is, all the lines along at least one path between the fault site and w have different v and v_f values.

In Figure 3.3 the **fault propagation** for fault $f = G_1$ s-a-1 occurs along the path (G_1, G_3, G_4). A line whose value for test t changes in the presence of the fault f is said to be **sensitised** to the fault f by the test t. A path composed of sensitised lines is called a **sensitised path**.

TG algorithms, referred to as *path-sensitisation* algorithms, operate based on these concepts. To find a test t that detects a fault G *s-a-v*, the fault must be *activated* by assigning appropriate input values to G so that $\bar{v}$ would normally be output. Then the resulting error must be *propagated* to a primary output (PO). Of course, these value settings on specified lines in the circuit must be *justified*, i.e. they must result from an assignment of primary input (PI) values. Finding such an assignment is a **line-justification** problem, which is solved by an *implicit enumeration* of all possible solutions. It consists of a recursive procedure in which the value of a gate output is justified by values of the gate inputs, until PIs are reached. For some values, some gate inputs may be left unspecified since it is enough that one of them takes the controlling value (absorbing element) of the gate. For instance, a 0-value at an *and*-gate's output is justified by a single 0-input, as shown in the example of Figure 3.4.

Figure 3.4. Justifying values for line k

3.3.3 Multi-valued Logics

To control error propagation, the composite logic values v/v_f must be considered. There are four such pairs of values, namely {0/0, 0/1, 1/0, 1/1}. Logic operations between composite logic values can be evaluated by composing the results for the fault-free and faulty circuits. Hence, a **4-valued logic** may be defined where values 1/0 and 0/1 represent errors and are denoted respectively by symbols D and $\bar{D}$ [Roth 1966] (see Figure 3.5). Values 0/0 and 1/1 are constants simply denoted by 0 and 1. As an example in the *or*-operation, $D + \bar{D} = 1/0 + 0/1 = (1+0) / (0+1) = 1/1 = 1$.

v/v_f	s	$\bar{s}$
0/0	0	1
1/1	1	0
1/0	D	$\bar{D}$
0/1	$\bar{D}$	D

AND	0	1	D	$\bar{D}$
0	0	0	0	0
1	0	1	D	$\bar{D}$
D	0	D	D	0
$\bar{D}$	0	$\bar{D}$	0	$\bar{D}$

OR	0	1	D	$\bar{D}$
0	0	1	D	$\bar{D}$
1	1	1	1	1
D	D	1	D	1
$\bar{D}$	$\bar{D}$	1	1	$\bar{D}$

Figure 3.5. 4-valued logic

In practice, logic operations with these values are defined by tables with an added fifth value (x) that denotes an unspecified composite value, that is, any value in the set $\{0,1,D,\overline{D}\}$. Figure 3.6 shows tables of basic operations over this **5-valued logic**. Each such logic operation corresponds to the two operations in the normal and faulty circuits since the 5 values are just an encoding of the two composite values.

NOT			AND	0	1	D	$\overline{D}$	x		OR	0	1	D	$\overline{D}$	x
0	1		**0**	0	0	0	0	0		**0**	0	1	D	$\overline{D}$	x
1	0		**1**	0	1	D	$\overline{D}$	x		**1**	1	1	1	1	1
D	$\overline{D}$		D	0	D	D	0	x		D	D	1	D	1	x
$\overline{D}$	D		$\overline{D}$	0	$\overline{D}$	0	$\overline{D}$	x		$\overline{D}$	$\overline{D}$	1	1	$\overline{D}$	x
x	x		x	0	x	x	x	x		x	x	1	x	x	x

Figure 3.6. 5-valued logic

Each assignment of a value to a line k may then *imply* other values, as exemplified in Figure 3.7. This may lead to an *inconsistency* during the search for a solution where decisions are made to justify a line or to propagate an error. If such a *conflict* occurs, there is a *failure* and *backtracking* is used to try a different choice. When all choices fail, there is no possible solution and the fault is said to be *undetectable*. A combinational circuit in which all stuck faults are detectable is said to be *irredundant*, otherwise, it is **redundant**.

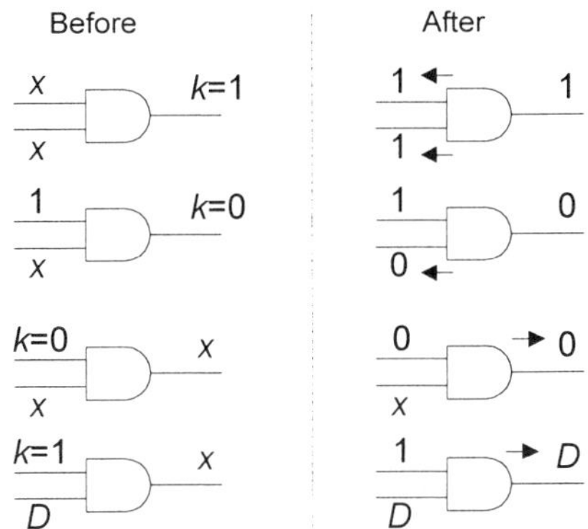

Figure 3.7. Some backward and forward implications using 5-valued logic

Path-sensitisation algorithms take a circuit with all lines initially unspecified and try to activate a given fault, propagate it to a PO and justify the decided line assignments. All value implications are propagated along the way according to the 5-valued logic, and if a conflict arises the last (incorrect) decision must be reversed by a backtracking mechanism.

In fanout-free circuits (see section 2.1) there is only one way to propagate an error to a PO, since any circuit signal line (including the fault activation site) has only one possible path to the circuit output. Also, line justification problems can be solved independently since the corresponding sets of PIs to assign are mutually disjoint.

For the general and usual case of circuits with fanout, the fault must still be activated leaving a line-justification problem, but there may be several options to propagate a fault. Once a propagation path is chosen, all that remains are line-justification problems. These problems, however, may no longer be independent with (reconvergent) fanout, and are a source for the possible conflicts. Also note that some faults may only be detectable with *multiple-path sensitisation*, where at least 2 reconvergent paths must be sensitised to propagate a fault to a PO, as is the case of fault G_1 s-a-1 in Figure 3.8.

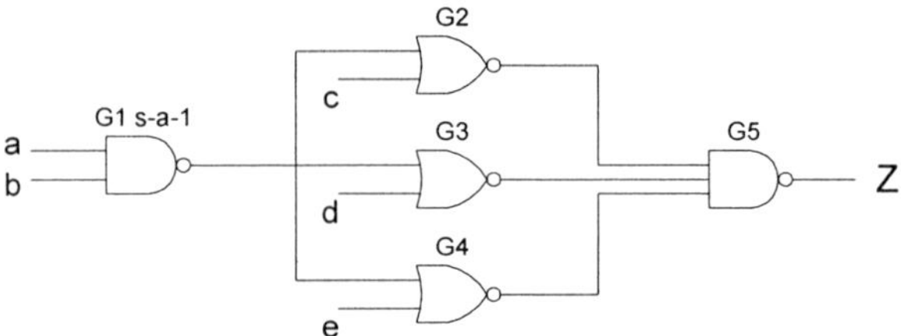

Figure 3.8. Target fault: G_1 s-a-1

Test vector *abcde*=11000 is the only one that detects G_1 s-a-1 by sensitising *all* three paths through G_2, G_3 and G_4 to yield a $\overline{D}$ at output Z (see Figure 3.9).

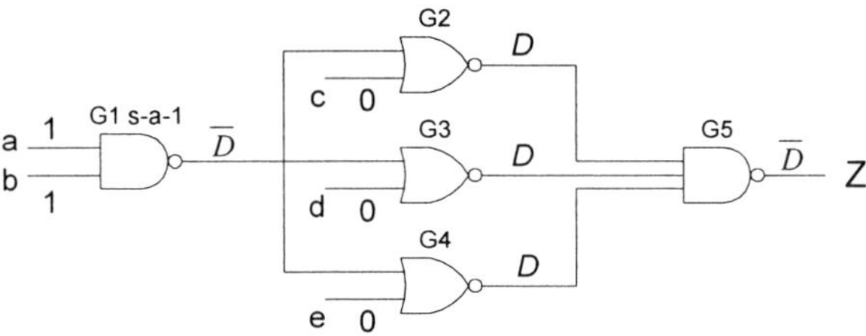

Figure 3.9. Multiple-path sensitisation

Such faults are considered rare in practical circuits [Cha *et al.* 1978]. So, to reduce computation time, some TG algorithms are often restricted to *single-path sensitisation*, as we will see in the next sections. This improves the speed of ATG systems, at the cost of possibly lowering fault coverage, by avoiding faults that are harder to detect. In the circuit of Figure 3.8, for the same fault G_1 s-a-1, after activating it by setting *a* and *b* to 1, if one first tries to propagate it through, say, G_2 (as in Figure 3.10) by setting *c* to 0, then the 2 other inputs of G_5 must be set to 1 to keep the (single) path sensitised. However, these 1-values (at the outputs of *nor*-gates G_3 and G_4) have no possible justification (i.e. they are inconsistent with the previous assignments) since an input is already $\overline{D}$ and neither value 0 nor 1 at *d* or *e* yields value 1 at the *nor*-gate since $nor(\overline{D}, 0) = D$ and $nor(\overline{D}, 1) = 0$.

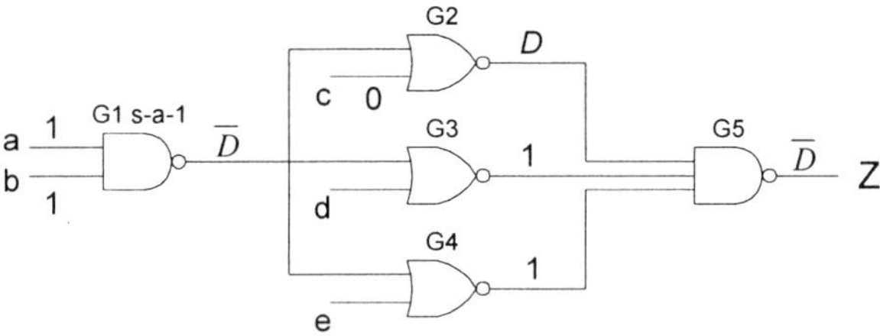

Figure 3.10. Impossible single-path sensitisation

To better deal with cases such as multiple-path sensitisation, a **9-valued logic** [Muth 1976] may be defined by adding four *partially specified composite values* to the 5-valued logic. The x value of the 5-valued logic is totally unspecified, that is, neither v nor v_f is known. For a partially specified composite value v/v_f, either v is binary and v_f is unknown (u) or vice versa. For example, $1/u$ represents both $1/0$ and $1/1$. This means that $1/u$ can be either D or 1. Table 3.2 shows the partially specified composite values and the sets of completely specified composite values they represent. The totally unspecified value x is u/u and represents the set $\{0,1,D,\overline{D}\}$.

Table 3.2. Partially specified composite values

v/v_f	values
$0/u$	$\{0,\overline{D}\}$
$1/u$	$\{D,1\}$
$u/0$	$\{0,D\}$
$u/1$	$\{\overline{D},1\}$

Table 3.3 below shows the 'not' table in this logic.

Table 3.3. NOT-operation in 9-valued logic

NOT	0/0	0/1	0/u	1/0	1/1	1/u	u/0	u/1	u/u
	1/1	1/0	1/u	0/1	0/0	0/u	u/1	u/0	u/u

This and other operations may be performed by separately taking values in the normal (v) and faulty circuits (v_f) and applying the same operation over the 3-valued logic. In fact, the 9-valued logic is just a double 3-valued logic ($3^2 = 9$). As an example in this logic, $D+x = 1/0 + u/u = (1+u) / (0+u) = 1/u$ which may be 1 or D, while in the 5-valued logic $D+x$ is simply x. Thus, the 9-valued logic provides more information, and a system using it may avoid backtracking in some cases, as we will see in the next section.

3.3.4 TG Specialised Algorithms

The **D-algorithm** [Roth 1966, Roth *et al.* 1967] is a classical TG algorithm over the 5-valued logic, where error propagation decisions (together with its implications) are given priority over justification problems, i.e. first the activated fault is propagated by successively choosing the gates of the sensitised path and appropriately setting its inputs, and only afterwards are these values justified.

Returning to the example of Figure 3.8, after activating fault G_1 s-a-1 by assigning 1 to both inputs a and b, the D-algorithm tries to propagate the error solely through G_2 (with c=0) by assigning 1 to the other inputs of G_5, which will lead to a contradiction, as we have already seen. Then it will try to propagate it through both G_2 and G_3, and only after this fails will it try and succeed to sensitise the three paths from G_1 (through G_2, G_3 and G_4) till Z.

The **9-V algorithm** [Cha *et al.* 1978] is an extension of the D-algorithm in that it uses the 9-valued logic. Since this logic provides more information than the 5-valued one, it may avoid backtracking in the case of multiple path sensitisation. In the example of Figure 3.8, the 9-V algorithm, after deciding propagating value D through G_2, will only have to partially specify the other inputs of G_5 to $1/u$, which forces d=0 and e=0 without backtracking, whereas the D-algorithm had to undo 2 wrong choices.

PODEM (Path-Oriented Decision Making) [Goel 1981] is an algorithm where a line-justification problem resulting from some decision (a value assignment made by PODEM to direct a path) is solved by a *direct search* process consisting only of PI assignments. An objective value v_k in a line k is backtraced to a PI i counting the inverters along the way (*xor*-gates must be translated into more basic gates) to yield the parity $p_{i,k}$ of inverters of the path so that the appropriate value $v_i = v_k \oplus p_{i,k}$ (likely to contribute to the objective) is assigned to PI i. This mapping of an objective to a PI assignment is repeated until the objective is achieved.

In the example of Figure 3.11, if the objective is to assign value v_f =1 to line f, PODEM may first backtrace it to PI *a*, which given value $v_a = v_f \oplus p_{af} = 1 \oplus (2\ mod\ 2) = 1 \oplus 0 = 1$ (objective value *xor* the parity of path inverters) does not yet imply the objective *f*=1. PODEM then backtraces again, this time to PI *b*, which given value $v_b = v_f \oplus p_{bf} = 1 \oplus (3\ mod\ 2) = 1 \oplus 1 = 0$ already achieves the objective *f*=1, since simulating the circuit with *ab*=10, *d* becomes 0, and, consequently, *f*=1. Thus, there are only *forward* implications (from inputs to outputs) which eliminates conflicts and may ease backtracking implementations since it can be done by *implicit simulation* rather than by an explicit save/restore process. PODEM is generally faster than the *D*-algorithm (experimental results also in [Goel 1981]).

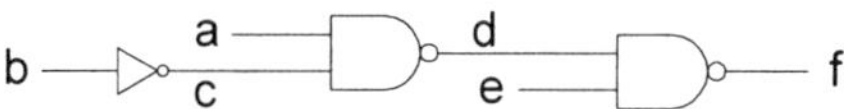

Figure 3.11. Objective: *f* = 1

The **FAN** (Fanout-Oriented TG) algorithm [Fujiwara and Shimono 1983] extends the backtracing concept of PODEM (which stops at PIs) with the possibility of stopping at internal lines that are called *head lines*, according to the following definitions.

- A *bound line* is a reachable line from (i.e. directly or indirectly fed by) at least one stem (see section 2.1). This means that any line with a possible path coming from a fanout point is a bound line.
- A line that is not bound is said to be *free*.
- A *head line* is a free line that directly feeds a bound line.

In the example circuit of Figure 3.12, lines *A* to *H* are free; *J*, *K* and *L* are bound lines; so, *G* and *H* are the head lines.

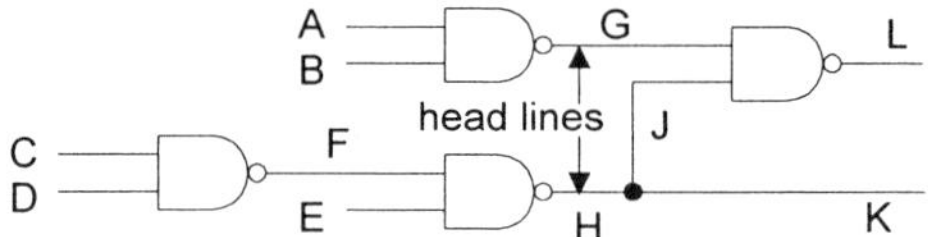

Figure 3.12. Head lines

Since the subcircuit feeding a head line *I* is fanout-free, a value of *I* can be justified without contradicting any other value previously assigned in the circuit. Thus backtracing can stop at *I*, and the problem of justifying the value of *I* can be postponed for the last stage of the TG algorithm. When the assignment of a value to a head line leads to a failure, PIs assignments of PODEM would only produce useless backtracking.

In addition, FAN also uses a *multiple-backtrace* procedure that attempts to simultaneously satisfy a set of objectives rather than one at a time. The objectives are the line-justification problems that come from the fault-activation and error-propagation problems. If, for instance, to activate a fault it is necessary that a set of gate inputs take a 0-value, PODEM would individually backtrace each objective, and a PI assignment allowing to satisfy one could preclude achieving other, forcing backtracking. FAN, in turn, may take all of the objectives together stopping backtracing at a highest-level stem and trying the most requested value there.

FAN has both forward and backward implications. It is more efficient than PODEM due primarily to the reduction in backtracking.

Other two algorithms exist that extend the concept of head lines by identifying higher-level lines whose values can be justified without conflicts. These lines are used to stop the backtracing process, but better results are not demonstrated.

In **TOPS** (Topological Search) [Kirkland and Mercer 1987], a *total reconvergence line* is the output *l* of a subcircuit *C* such that all paths between any line in *C* and any PO go through *l*. Note that the basic gates implementing the *xor* function may constitute such a subcircuit and its output is a total reconvergence line. Another algorithm, **FAST** (Fault-oriented Algorithm for Sensitized-path Testing) [Abramovici et al. 1986a], based on an analysis that is both topological and functional, uses a line *l* as a *backtrace-stop line* for value *v*, if the assignment *l=v* can be justified without conflicts.

While the worst-case (observed mainly for undetectable faults [Cha *et al.* 1978]) complexity of all these specialised TG algorithms remains exponential due to the possible number of incorrect decisions, on the average-case they are able to run on acceptable time, as is typical in many NP problems.

3.3.4.1 Extra Techniques

Improvements to TG algorithms such as *global implications* of **SOCRATES** [Schulz *et al.* 1988, Schulz and Auth 1989] may achieve a better efficiency. In the circuits of Figure 3.13, the value of *Z* always implies a value for *B*. In (a), when *Z* is assigned 1, no local implications can be made, but all possible justifications for *Z*=1 necessarily have *B*=1. SOCRATES "learns" this global implication in a preprocessing phase by simulating *B*=0 and verifying that it implies *Z*=0. Then $\overline{Z=0}$ implies $\overline{B=0}$, that is, *Z*=1 implies *B*=1, independently of other values in the circuit, being thus called *static learning*.

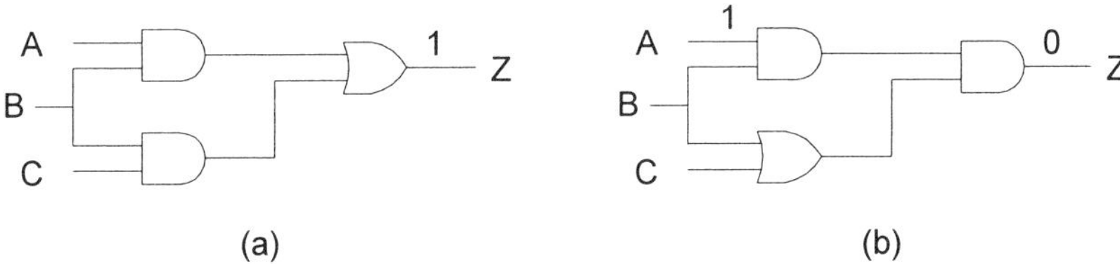

Figure 3.13. Global implications: the value of *Z* determines the value of *B*

Dynamic learning is also performed by SOCRATES to determine global implications enabled by previously assigned values. For example, in Figure 3.13 (b), when *A* is already 1, assigning 0 to Z implies *B*=0 because if *B* were 1, *F* would also be 1 in that case.

Efficient TG for large circuits is, in fact, an important practical problem, and the consensual opinion of the scientific community is that it requires hardware support or Artificial Intelligence (AI) techniques, the two current main research directions in this area.

As we will see in the next section, constraint reasoning can be efficiently applied to the TG problem, naturally incorporating many of the discussed techniques while other improvements may easily be introduced.

3.4 Constraint Reasoning

Logic Programming (LP) is a very natural approach for TG due to its embedded backtracking mechanism, and *Constraint Logic Programming* (CLP) adds expressive power and efficient execution to the declarative nature of LP. Circuit gates may be represented in CLP as constraints that model their functional behaviour, and decisions for the search procedure expressed by simple logic disjunctions. On providing the target fault, the CLP system solves the constraints, implicitly propagating the implications and, if there is a solution, generates the test pattern.

3.4.1 CLP(B)

In a CLP(B) system (where B stands for Boolean domains), one can easily model a digital circuit by modelling in turn each logical gate by means of a Boolean constraint on the gate's inputs and output. A CLP(B) system may be complete by using Boolean unification, in which case it is referred to as *symbolic*, or it may be incomplete be handling variables as finite domain ($\{0,1\}$) variables as in a CLP(FD) system. In this section we discuss these two types of Boolean solvers.

3.4.1.1 CLP(B) - Symbolic

A circuit represented as a set of Boolean constraints can be solved using a symbolic Boolean approach [Büttner and Simonis 1987], thus corresponding to an algebraic model. For instance, in Figure 3.13 (a) circuit output Z is given by $Z = (A+C).B$. Hence, when $Z=1$, we have $(A+C).B = 1 \Leftrightarrow A+C = 1 \wedge B=1$ (with Boolean unification, this may be represented as $\{A/\ (C \oplus 1)+A',\ B/1\}$). We may thus infer the global implication $Z=1 \Rightarrow B=1$. Nevertheless, such symbolic reasoning implicitly computes all solutions, thus turning a problem such as TG (where the usual goal is to find a solution) into the NP-hard complexity category.

3.4.1.2 CLP(B) – Finite Domain

Boolean constraints can be solved more efficiently by a number of incomplete Boolean solvers, obtained as a specialisation of a finite domains solver [Codognet and Diaz 1997].

In [Simonis 1989], the author suggests a technique to generate test patterns for SSFs, implemented in the CLP system CHIP [Dincbas *et al.* 1988]. It implicitly adopts the 5-valued logic of Figure 3.6 using Boolean domain variables for 0- and 1-values, and symbolic values d and *dnot* for **d-signals** D and $\overline{D}$. An uninstantiated Boolean domain variable may represent the unspecified value x. The algorithm can be classified as topological, in that it *activates* the fault, *propagates* it to the output and *justifies* the choices made.

To find a test pattern for the given fault, it is necessary to traverse the circuit in increasing level order (from inputs to outputs) posting the Boolean gates constraints with the Boolean variables corresponding to their connections. When the gate under test is reached, the fault is *activated* by appropriately setting the gate Boolean inputs (all of them or just one by means of a disjunction) so that the gate outputs a d-signal. When a d-signal reaches the input of a gate, its *propagation* is always forced by CHIP *demons* [Davis 1984, Simonis and Dincbas 1987], which avoid its masking by imposing some constraints on the other inputs. Demons are constraints that behave in a data-driven way, by means of a set of rules that describe the action to perform when the arguments satisfy some condition (e.g. a ground argument with some value). For example, when an input of an *or*-gate is symbolic value d, then the other inputs are set to 0 and value d is output. D-signals are thus always explicitly set, leaving only Boolean variables in the circuit.

If a d-signal reaches an output bit, the sensitised path must be *justified*. The resulting Boolean *Constraint Satisfaction Problem* (CSP) must be satisfied by some labelling of the remaining variables to find a proper test pattern. In spite of using a 5-valued logic (4 explicit values plus the unspecified Boolean domain variable), the CSP that is generated for a chosen sensitised path (with the symbolic d-signals) is purely Boolean. There may thus be many possible different Boolean CSPs for some TG problem, according to the choices made regarding the sensitised path. The system must find a solution to one such CSP to solve the problem.

By always enforcing fault propagation, Simonis approach may overlook some solutions. In Figure 3.14, for example, the value to assign to a must be chosen. In this example, the d-signal must pass solely through the *not*-gate and be masked at the *or*-gate by $a=1$ (Figure 3.15 (b)). By enforcing $a=0$ (to propagate the d-signal), Simonis does not allow it to reach z (Figure 3.15 (a)), since the two opposing d-signals will meet at the *and*-gate thus annihilating themselves.

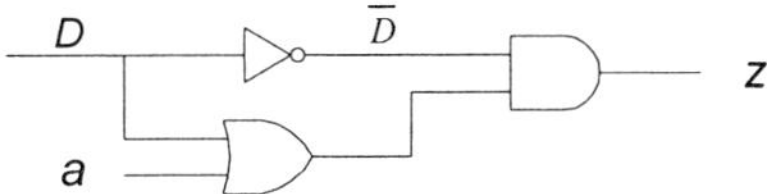

Figure 3.14. How to assign *a* to have a *d*-signal at z?

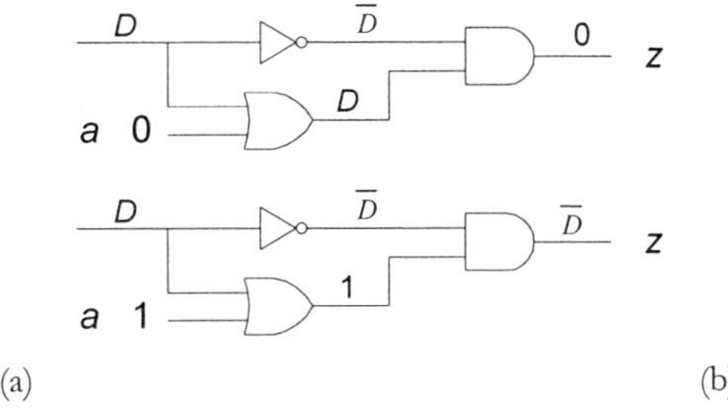

(a)   (b)

Figure 3.15. To have a *d*-signal at z, D must be masked at the *or*-gate by $a{=}1$

Replacing the demons by disjunctions to allow more solutions should solve this problem, i.e. instead of forcing propagation (as in the *or*-gate by $a{=}0$), the alternative hypothesis $a{=}1$ should also be allowed, although locally masking the *d*-signal. In fact, globally, it is the only way that enables the *d*-signal to reach a PO.

In [Azevedo and Barahona 1998] we present a diagnostic tool able to generate these test patterns with a constraint solver over Boolean variables but extending the scope to Multiple Stuck Faults (MSFs). Gates either propagate or mask *d*-signals, but preference (in the sense of priority in search) is given to propagation. To avoid carrying on a computation when no more *d*-signals are present, it is sufficient to maintain one counter for the current number of *d*-signals present in the circuit at any step of the computation. After fault-activation, the program may backtrack as soon as the counter is null.

If this heuristic procedure works well in many situations, in others the number of choice points created is too large. Moreover, backtracking on a choice point is caused by the failure of solving a possibly large Boolean CSP. This makes critical the heuristics made at each choice point. Additionally, there can be obvious optimisations that the system will not find, as shown in the circuit of Figure 3.16.

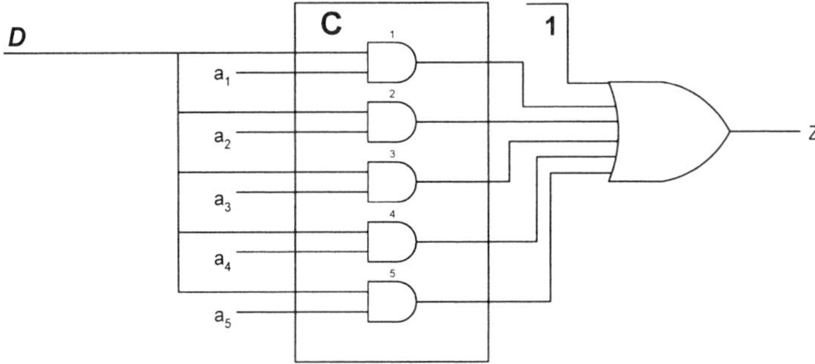

Figure 3.16. Drawback of creating choice points

In this very simplified example, D can never propagate to the output Z, yet 2^5 decisions (on a_i

values) are made to reach this conclusion. If sub-circuit C contained more choice points, execution time would increase exponentially. This is a typical situation with some circuits (e.g. *c432* of [ISCAS 1985]), which, even if relatively small, create a number of difficulties hard to handle. Note that the above mentioned TG algorithms also face this problem. Applying the single-path sensitisation restriction here may largely compensate in terms of time by reducing backtracking, but, as mentioned, at the possible cost of not finding the solution even if one exists.

3.4.2 CLP(FD)

In [Simonis 1992], the author presents another TG version for CHIP, where fault propagation is not always enforced, and where two extra symbolic values: e and *enot* are introduced. Such e-signals have the same meaning of d-signals but can be masked. When a d-signal (d / *dnot*) is at a fanout point, it may "choose" one and only one line to follow, while the other lines take the e-signal (e / *enot*). The goal is still to have a d-signal at the output, therefore it cannot be masked and, at the end, there will be a single-path with d-signals, although there can be many with e-signals reconverging to it, thus allowing multiple-path sensitisation.

In addition, this version extends the domain of the constraints from Boolean to finite domains of 6 values $\{0, 1, d, dnot, e, enot\}$ in a CLP(FD) approach which allows the *delaying* of some choices of values in the circuit. (In practice, the variables to label in the CSP have only 4 possible values, since a path with d-signals is chosen as before.)

With such solver, Simonis developed an ATG system (section 3.2), and applied it to the ISCAS circuits described in section 2.4. Table 3.4 reproduces the results obtained on a SUN 3/260 taken from [Simonis 1992], where 'Red.', 'Ab.' and '#' represent respectively, the number of redundant faults, aborted faults and the total number of tests obtained (i.e. the cardinality of the circuit test set). The total time needed is shown in seconds, as well as the average time per test generation (T/TG) attempted.

Table 3.4. Results of Simonis' ATG system

	Red.	Ab.	#	Time	T/TG
c432	1	3	68	180	2.5
c499	8	0	62	147	2.1
c880	0	0	74	225	3.0
c1355	8	2	92	525	5.1
c1908	5	5	124	920	6.9
c2670	97	41	105	3285	13.5
c3540	127	25	175	4208	12.9
c5315	59	22	141	3821	17.2
c6288	34	0	37	2358	33.2
c7552	88	122	281	9924	20.2

As with other ATG systems, emphasis is given to finding a small test set while assuring high fault coverage. Test sets may be much smaller than the whole set of faults since a test may detect many faults, as previously explained. The general idea of the ATG system is to generate a test pattern for one of the still undetected faults and, if successful, perform fault simulation in the circuit with the obtained test to check what other faults are also detected (such simulation is discussed in Chapter 5). If it is impossible to generate a test, then the fault is redundant. If execution terminates by reaching a user-specified limit, without finding a test, then the fault is aborted. In both cases, the system moves on to the following undetected fault, until all faults have been considered.

3.4.2.1 Our Solver

In [Azevedo and Barahona 2000c], in addition to our use of an extended logic for related diagnostic problems (see next chapters), we show that TG for SSFs or MSFs can be solved by constraints over the 4-valued logic with the semantics specified in the tables of Figure 3.5, thus avoiding the choice points at fanouts with a *d*-signal.

These logic tables are the ones of 5-valued logic with removed *x* entries. Finite domain (FD) variables have just 4 possible ground values. Uninstantiated variables are unspecified (*x*) values.

In a CLP model of a digital circuit, the domain variables represent the values of the inputs and outputs of gates. Variable-sharing models a connection among gates (i.e. if gate g_1 directly feeds gate g_2, then the output variable of g_1 is an input variable of g_2), which implies that a fanout is topologically a point with a single value. To model all possible stuck lines, fanout lines are thus replaced by buffers in the model, so only stuck gates have to be considered (PIs are also treated as buffers) by means of the corresponding S-buffers (see section 2.5).

A circuit is then basically modelled by the set of constraints corresponding to its gates. Each original circuit line contributes with one constraint, and an additional constraint models a target fault. Although the final model is independent of the order of posted constraints, it is convenient to keep the order of constraints from PIs to POs due to domain declarations and propagations (for example, all variables appearing before an S-buffer are already constrained to the Boolean domain {0,1} instead of the full 4-valued domain). This model is free of choice points regarding sensitised paths since no decisions are made along the way, i.e. there is no commitment to a particular value in a signal line or to some sensitised path (such decisions are *delayed* by using constraints over the 4 logic values). Hence, there is only one CSP for a TG problem. Either this CSP is solved and we find a test pattern, or it fails and the fault is redundant.

This approach makes it quite straightforward to generalise TG on SSFs to multiple faults. In TG for multiple faults, we only have to take into account that S-buffers may already have a *d*-signal (from another fault) at its input. All 4 logic values (0, 1, D, $\overline{D}$) are thus possible inputs for S-buffers.

Let S-buffer/0 (/1) denote an S-buffer s-a-0 (s-a-1), i.e. a buffer that in the faulty circuit model always outputs 0 (1). S-buffers behaviour in the 4-valued logic can also by determined by separately examining the results for the normal and faulty circuits, and combining them as in Table 3.5.

Table 3.5. S-buffers logic table

Input	S-buffer/0 output	S-buffer/1 output
$0 = 0/0$	$0/0 = 0$	$0/1 = \overline{D}$
$1 = 1/1$	$1/0 = D$	$1/1 = 1$
$D = 1/0$	$1/0 = D$	$1/1 = 1$
$\overline{D} = 0/1$	$0/0 = 0$	$0/1 = \overline{D}$

Note that in case of multiple faults it is not necessary to activate them all. One fault activation may be enough, as long as it is propagated to a PO. Consequently, a fault is not immediately activated. Rather, activation or any other decision (commitment subject to failure and backtracking) are delayed until all constraints have been posted and propagated.

With such small domains, arc consistency (section 1.2) may be applied effectively. Thus, an S-buffer/0 (/1) with a Boolean domain variable input, outputs a variable with domain {0,D} ({1, $\overline{D}$}). This is similar to the partially specified composite values of the 9-V algorithm. Domain variables are still unspecified but may carry more information than the partially specified values of 9-V, since all combinations of the 4 values are possible in a variable domain. The different possible domains for a variable correspond to $\mathcal{P}(\{0, 1, D, \overline{D}\})$ (the powerset of the full domain), excluding the empty set ({} corresponds to a failure) and singletons (which are the 4 ground values). There are $\binom{4}{4}+\binom{4}{3}+\binom{4}{2} = 1+4+6 = 11$ such combinations, which added to the 4 ground values yields what can be seen as an implicit 15-valued logic (or the full $2^4 = 16$ valued logic with {} as the absorbing element of all logic operations). Table 3.6 shows the logic table of an *and* operation, where d and n stand, for space reasons, for D and $\overline{D}$, and where each table entry corresponds to the domain of the possible input values 'anded'). Thus, a full domain ($\{0, 1, D, \overline{D}\}$) variable is the totally unspecified logic value, while other (10) reduced domain variables can be seen as partially unspecified values.

Table 3.6. Implicit 16-valued logic conjunction

And	{}	0	1	d	n	01	0d	0n	1d	1n	dn	01d	01n	0dn	1dn	01dn
{}	{}	{}	{}	{}	{}	{}	{}	{}	{}	{}	{}	{}	{}	{}	{}	{}
0	{}	0	0	0	0	0	0	0	0	0	0	0	0	0	0	0
1	{}	0	1	d	n	01	0d	0n	1d	1n	dn	01d	01n	0dn	1dn	01dn
d	{}	0	d	d	0	0d	0d	0	d	0d	0d	0d	0d	0d	0d	0d
n	{}	0	n	0	n	0n	0	0n	0n	n	0n	0n	0n	0n	0n	0n
01	{}	0	01	0d	0n	01	0d	0n	01d	01n	0dn	01d	01n	0dn	01dn	01dn
0d	{}	0	0d	0d	0	0d	0d	0	0d	0d	0d	0d	0d	0d	0d	0d
0n	{}	0	0n	0	0n	0n	0	0n	0n	0n	0n	0n	0n	0n	0n	0n
1d	{}	0	1d	d	0n	01d	0d	0n	1d	01dn	0dn	01d	01dn	0dn	01dn	01dn
1n	{}	0	1n	0d	n	01n	0d	0n	01dn	1n	0dn	01dn	01n	0dn	01dn	01dn
dn	{}	0	dn	0d	0n	0dn	0d	0n	0dn	0dm	dn	0dn	0dn	0dn	0dn	0dn
01d	{}	0	01d	0d	0n	01d	0d	0n	01d	01dn	0dn	01d	01dn	0dn	01dn	01dn
01n	{}	0	01n	0d	0n	01n	0d	0n	01dn	01n	0dn	01dn	01n	0dn	01dn	01dn
0dn	{}	0	0dn	0d	0n	0dn	0d	0n	0dn	0dn	0dn	0dn	0dn	0dn	0dn	0dn
1dn	{}	0	1dn	0d	0n	01dn	0d	0n	01dn	01dn	0dn	01dn	01dn	0dn	1dn	01dn
01dn	{}	0	01dn	0d	0n	01dn	0d	0n	01dn	01dn	0dn	01dn	01dn	0dn	01dn	01dn

Table 3.7 shows some examples of constraint propagation with our CLP(*FD*) solver with the relevant variable domains for the 4-valued logic. A constraint on a logic operation updates the domains of its arguments (variables) making them arc-consistent (hyper-arc-consistent, actually, since a constraint may involve 3 variables). The constraint may be subsequently removed from the store if the performed propagation implies its satisfiability, i.e. it is subsumed by the new constraint store, or, in other words, any solution to the remaining CSP is a solution to the constraint, so it may be removed with no loss of information. Removable constraints are marked with a star (*) in the table.

Table 3.7. Some constraint propagation examples in 4-valued logic

Initial Domains			Constraint	Final Domains		
X	*Y*	*Z*		*X*	*Y*	*Z*
$\{0,1\}$	$\{0,D\}$		$Z=X{\vee}Y$			$\{0,1,D\}$
$\{0,1,D\}$	$\{1,\overline{D}\}$		$Z=X{\wedge}Y$			$\{0,1,D,\overline{D}\}$
$\{0,D\}$	$\overline{D}$		$Z=X{\wedge}Y$ *			0
$\{0,1,D\}$			$Z=\overline{X}$			$\{0,1,\overline{D}\}$
$\{1,D\}$			$Z=\text{s\_buff}/0(X)$ *			D
$\{1,D\}$	$\{1,D\}$		$Z=X{\oplus}Y$			$\{0,\overline{D}\}$
		$\{0,D\}$	$Z=X{\vee}Y$	$\{0,D\}$	$\{0,D\}$	
$\{1,D,\overline{D}\}$	$\{0,1,\overline{D}\}$	$\{0,D\}$	$Z=X{\vee}Y$ *	D	0	D

The basic propagation rule to maintain arc-consistency is that when a variable is updated, other variables sharing a constraint are updated accordingly, i.e. impossible values are removed from their domains. This can be synthesised with the rule below, where $c(X,Y,Z)$ is a constraint over variables X, Y, Z, and D_X, D_Y, D_Z represent their domains (operator '::' is used for domain assignment):

$$(\text{X: changed}) \quad \frac{D_Y'=\{y\in D_Y, \exists_{x\in D_X, z\in D_Z}, c(x,y,z)\}, D_Z'=\{z\in D_Z, \exists_{x\in D_X, y\in D_Y}, c(x,y,z)\}}{\{c(X,Y,Z)\}\mapsto\{Y::D_Y', Z::D_Z', c(X,Y,Z)\}}$$

If all combinations of values from the domains of the variables of a constraint are possible (i.e. satisfy the constraint), then the constraint is removed from the store.

Our solver ensures full arc-consistency on the constraint network in a preprocessing stage (setting up circuit constraints), and maintains partial arc-consistency during search. While binary constraints such as *not*-gates and S-buffers are always kept fully arc-consistent, other constraints such as *xor*-gate (involving 3 variables) and *and*-gate (which may have an arbitrary number of inputs) are kept arc-consistent only if all their variables are Boolean; otherwise propagation is delayed until one constraint variable becomes instantiated (as in forward checking [Haralick and Elliott 1980]).

When triggering a constraint due to the instantiation of a variable, we again enforce arc-consistency on it (in a similar but weaker fashion to MAC – Maintaining Arc Consistency [Sabin and Freuder 1994]) and, if the remaining variables are only two or all Boolean, then the constraint is made fully arc-consistent; otherwise propagation is delayed until another variable becomes instantiated.

3.4.2.2 *Advantages of Constraint Propagation*

After successful propagation of all circuit constraints, search for a solution may start if constraints remain on the store. To ensure that a solution may still be possible, there must be at least one output bit with some *d*-signal in its domain. Otherwise, we may already conclude that there is no solution, which proves that the target fault is undetectable.

Let us analyse again the circuit of Figure 3.16. Here, we immediately conclude that the output can have no *d*-signal, since its value is 1. This conclusion requires no backtracking at all since no decisions had to be made while setting the circuit gate constraints. In contrast, all the other algorithms discussed above, which explicitly try to set sensitised paths, could only reach such a conclusion after exhaustively covering the search space that resulted from the created choice points.

In the circuit of Figure 3.17, where a *d*-signal is needed at the output, no matter through which path(s) the error propagates, a global implication can be made, namely that z takes value D and k value 1 [Akers 1976, Fujiwara and Shimono 1983]. While other algorithms would have to explicitly consider such cases to improve efficiency, the propagation of constraints makes these situations quite naturally dealt with.

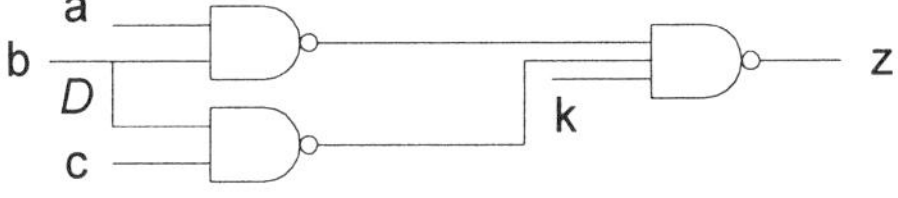

Figure 3.17. Global implication: D must pass through z

Other situations such as that shown in Figure 3.18 lead to trivial failure of the constraint system. Other algorithms need to use a look-ahead technique by verifying that an error can only propagate to a PO if there exists at least one path whose lines from the error to the PO all have value *x* [Goel 1981].

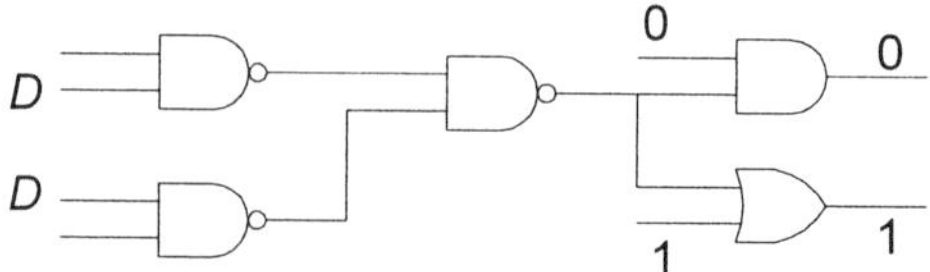

Figure 3.18. Impossible error propagation

If the *d*-signal is indeed possible at the output, it must come from a source S-buffer. The cardinality operator [Van Hentenryck and Deville 1991] is generally used in CP to impose such disjunction of constraints (i.e. that one of them outputs a *d*-signal). If only one S-buffer exists (the case of an SSF), or only one is able to activate the fault, then the constraint solver automatically activates it and performs the appropriate constraint propagations (implications).

The number (parity) of inverters between the S-buffer(s) and the output does not have to be explicitly checked (as done in PODEM) since inverter constraint propagations are sufficient to disallow impossible *d*-signals. In Figure 3.19 we show some examples where variable domains are represented in brackets. Due to the inverters we know that z can only assume values 0 or D (never $\overline{D}$). If, for instance, it is 'anded' with $\overline{D}$ the result will simply be 0.

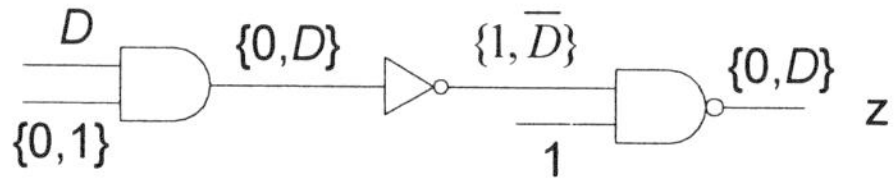

Figure 3.19. Inverter constraint propagations

3.5 Heuristics

A search process is guided by decisions that aim at solving the problem faster or finding a "better" solution according to some cost function. In basic TG any solution is a good one, but one may prefer more unspecified input bits or some input bits set to 0 or 1, for purposes such as test set compaction. Heuristics are criteria to select the most "promising" choices in decision-making. To solve a CSP, there are basically two kinds of decisions:

1. Which variable should be instantiated first, that is, which sub-problem to try first ?
2. Which value should be assigned first to the chosen variable ?

Normally, harder sub-problems are tried first to avoid wasting time in solving easy sub-problems only to find out later (possibly after much backtracking) that an impossible difficult problem still requires to be solved. For problems where variables play similar roles, the first-fail heuristic [Haralick and Elliott 1980] consists of selecting the variable with the smaller domain, and is usually very effective. The constrainedness of a variable may also be considered: more constrained (i.e. with more constraints over it) variables are selected first. The rationale for this preference is that such variables are harder to instantiate and doing it first will propagate much information to other variables along the constraint network, thus greatly reducing search space.

A variable may also be selected first because it plays a crucial role in the problem (e.g. by being strongly connected to its goal). This is the case in TG where the goal is to find a PO where values differ for the normal and the faulty circuits, that is, to find a PO with a *d*-value as a result of an activated and propagated fault. Consequently, POs that may satisfy the final goal are strong candidates for starting the enumeration of variables (lines). Constraint propagation eases the task of selecting a PO since only a fraction of them contain *d*-signals in the domain. So, we may start by selecting any *sensitisable* PO and then justify it by labelling only its relevant circuit inputs (in the transitive fanin).

Solving such a disjunctive constraint (find at least 1 sensitised PO) is often improved by using the cardinality constraint, since its least commitment nature delays choices and avoids many wrong decisions. Unfortunately, this approach also delays the knowledge about relevant inputs, since any of the possible POs may be sensitised. Hence, all inputs with possible paths to those POs involved in the cardinality constraint are potentially relevant. In general, more inputs (i.e. variables) have to be labelled, which decreases efficiency.

To illustrate these comments, we show in Table 3.8 the results obtained by labelling first those input variables leading to the S-buffer and then those remaining which lead to the candidate POs. Each TG attempt was given a maximum time limit of 2 minutes, after which it would abort. Since efficiency of TG algorithms are generally assessed from results of an embedding ATG system, we implemented one in ECLiPSe Prolog 5.1 (Tcl/Tk version) under Windows 2000 on a Pentium 4, 1.7 GHz, 512 Mb RAM.

Table 3.8. ATG results with cardinality constraint

	Red.	Ab.	#	Time	T/TG
c432	1	95	76	11500	66.9
c499	0	16	86	1993	19.5
c880	0	51	95	6447	44.2
c1355	0	49	157	6308	30.6
c1908	8	41	181	5743	25.0
c2670	11	480	125	57672	93.6
c3540	15	1002	150	117113	100.4
c5315	23	820	162	97675	97.2
c6288	18	320	403	47127	63.6
c7552	8	839	328	103993	88.5

As one can see from the large number of aborted faults, heuristics for selecting POs are crucial to handle this problem. The least commitment philosophy of constraint solving did not allow us to restrict the number of variables to label, which led to quite poor execution times. Choosing a specific PO to sensitise can drastically reduce the variables to label.

When choosing a value for a variable, preference is given to values that are more likely to succeed, or that may produce a better (preferred) final solution, or a combination of these. For the PO, any d-signal will satisfy the goal, so they are preferred. Since, at this point there is no reason to choose a specific D or $\overline{D}$ (either will do), we may simply remove the Boolean values from the variable domain. If only one of these d-signals is possible, the variable is instantiated with it; otherwise the choice is delayed until another stage or until a value is implied by subsequent choices.

If a specific PO z is selected to be sensitised, then z is the highest-level line of at least one sensitised path propagated from a fault. Hence, one of the inputs of the gate whose output is z is also sensitised. Again, only a fraction of these are possible and the process of choosing one and forbidding Boolean values may be repeated. This recursive process eventually reaches a target fault (an S-buffer) that is then activated if it was not already activated (in the case of a SSF, the single fault can be immediately activated when setting up circuit constraints, since it is the only S-buffer from which it can be propagated). Choosing a path and reaching an S-buffer determines specific D or $\overline{D}$ values that may have been left to decide along this path, i.e. variables that had D and $\overline{D}$ in the domain become instantiated, since a specific d-signal is generated when the fault is activated and for each gate in the chosen path, it either is negated (in *not-*, *nand-* and *nor-*gates) or remains the same value (in $\{buffer, and, or\}$). Hence, the sensitised PO is also instantiated. (This may not be true if there were *xor*-gates involved in the chosen path, in which case even a specific d-signal input may be insufficient to instantiate the sensitised output since a Boolean domain variable on the other *xor* input could "pass" or invert the *d*-signal according to that final Boolean value, 0 or 1, respectively.) Anyway, a sensitised path was chosen and must be subsequently justified.

These line-justification problems from the S-buffer input to the PO z are solved in a FAN-like way by labelling the relevant PIs stopping at head-lines where possible. The reason for this "upwards" labelling starting at an S-buffer is that while moving from line l_1 to the higher-level line l_2, the relevant PIs increase in number, i.e. the relevant PIs of l_1 are a subset of those of l_2. Hence, at each stage, labelling PIs or internal lines for a line-justification problem is also relevant to the next ones.

If no solution is found in this process, then there is no global solution and the fault is redundant. Otherwise, the PIs not involved in the enumeration may be left unspecified since their values will not interfere in the solution. Justification of internal head-lines may also lead to some unspecified PIs since the justification of the corresponding fanout-free sub-circuit is done "downwards" by implicit enumeration as in Figure 3.4. The final solution is the desired test pattern.

3.5.1 Discussion and Potential Improvements

The unspecified inputs of a generated test pattern may, nevertheless, be made specified so as to possibly detect more faults. The subsequent simulation may then discard several more faults (a process referred to as *fault dropping*) that will not need specific TG (since they are already detected), thus possibly reducing the final test set, total execution time and number of aborted faults. This is where specialised heuristics can really make a difference in the final ATG output. Since we are mainly interested in simple TG for some diagnosis, we stick to labelling the remaining input variables (i.e. the yet unspecified inputs) in a random way.

In this CLP implementation, path sensitisation also requires some choices to be made but one does not need to stick to some chosen path and afterwards solve the justifications problems. On the contrary, this approach delays commitments as much as possible, and constraint propagation is performed among small choices made to reduce search space so as to try to find a path. Only after no more propagation is possible, some choice is made. Hence, there is a better integration between commitment and justification.

Additionally, other heuristics in TG can be used applying other cost functions such as the *observability* and *controllability* of lines and values [Abramovici *et al.* 1990], as Simonis does. These cost functions help choosing lines to justify first or what values to try first, by providing *relative measures* of their "difficulty". In general, cost functions are computed by a pre-processing step and can be distance-based, recursive and fanout based. But no matter the method, cost functions should show PIs as the easiest signals to control and POs the easiest to observe. The measures should require significantly less computational effort than the TG process; otherwise it is not worth it. Anyway, two different circuits may have contrasting results with two different heuristics. This is a natural consequence of heuristics since they are based on approximate estimations that make them circuit-dependent. To compensate for this effect, it is possible to switch among cost functions during TG [Chandra and Patel 1989]. Although these specialised heuristics could really make a difference, we considered them out of the scope of this thesis and decided not to use them.

Also, performing some form of intelligent backtracking (e.g. dependency-directed backtracking [Stallman and Sussman 1977], backjumping [Gaschnig 1979], *k*-order learning [Dechter 1990] or dynamic backtracking [Ginsberg 1993]) could solve line-justification problems more efficiently. We relied on the traditional chronological backtracking of Prolog for labelling a set of PIs (or head-lines), which can be computationally heavy for a large set of variables if the reason for a detected conflict lies in some early labelling choice among those bits. Intelligent backtracking could possibly detect that reason and save a lot of time by directly undoing that incorrect choice.

3.6 Iterative Time-Bounded Search

The goal of having a *d*-signal in one of the output bits of a digital circuit is a typical disjunctive constraint (it is either in the first output bit, or in the second, or the third, …). As observed earlier, the classical technique to efficiently handle such constraints consists of delaying the choice of the alternatives until a final labelling eventually makes a commitment to one of the disjuncts. Such technique, using the cardinality operator, is usually much more efficient than traditional depth-first search, since making an early mistake can be very costly to undo.

However, the *least commitment* strategy has a price: the problem is kept less constrained than that resulting from making an early commitment. In particular, less propagation is usually possible, or only weaker heuristics may be used in the less constrained problem.

The problem with early commitment is thus the difficulty in undoing the wrong choices. But often, there is a large difference in complexity of solving the different problems that result from making each of the possible choices. Some of these, might also be much more easy to solve than the whole problem. If this is the case, it pays off to spend some time trying to solve each of the sub-problems in a round-robin discipline. To make the strategy complete, the time allowed for each of the tries must be increased in subsequent round-robin turns. This is the basic idea behind the **Iterative Time-Bounded Search** (ITBS) that we propose for this kind of problems and have applied for the specific case of TG.

The method may be described as follows. If the initial problem is composed of a disjunction with k disjuncts, ITBS assigns a time limit T to solve each of the sub-problems, thus making a limited commitment to each disjunct. If all sub-problems are aborted due to exceeding time limit T, this limit is doubled (more generally multiplied by some factor f). The time is thus increased in each of the rounds, until a solution is found.

ITBS thus shares the underlying idea of iterative broadening and iterative deepening and other techniques (e.g. [Ginsberg and Harvey 1990, Harvey and Ginsberg 1995, Meseguer 1997, Walsh 1997]) to overcome the problems of depth-first search, by searching side-branches before fully exploring a previously selected branch of the search space.

If a solution is found in round r $(r \geq 1)$, the ITBS worst execution time is $A = k*(T+T*f+\ldots+T*f^{r-1})$, occurring if the problem is solved with the last choice. Compared with the best execution time achieved when committing to the right choice but with no time limit, i.e. $B=T*f^{r-1}$, ITBS is penalised by a factor of

$$A / B = [k*T*(f^r-1)/(f-1)] / (T*f^{r-1}) \approx k*f / (f-1) \approx k$$

Hence the penalty for committing to the wrong PO is linear on the number of relevant POs. The ITBS strategy pays off whenever there is some heuristic that solves one sub-problem much faster than solving the whole problem with least commitment. This is, of course, problem dependent.

We applied ITBS to the TG problem considering two different strategies for path sensitisation:

1- Each disjunct (try) consists of sensitising (constraining) a PO by giving it the domain $\{D,\overline{D}\}$; the choice of path then proceeds by similarly constraining gate signals downwards to an S-buffer.

2- Each disjunct consists of sensitising (instantiating) a PO with a specific *d*-value; the choice of path then proceeds by instantiating *d*-values to gate signals downwards to an S-buffer.

In both cases, we gave each try an initial time limit T of 2 seconds to solve the whole problem.

This limit was subsequently doubled (f=2) on two more rounds (maximum time limit of 2*2*2=8 seconds). As explained, for each chosen PO (or PO value), a sensitised path is found and relevant inputs (from S-buffer to PO) are labelled in the given order simply trying first value 0 and then value 1. When a solution is found, the TG problem is solved.

For the first strategy (constrain path to be sensitised), if some try over PO z is proven to be impossible, then it is marked as such and subsequent tries over remaining POs may add the constraint that PO z is Boolean (not sensitised). For strategy number 2 (specified sensitised path), if $z = D\,(\overline{D})$ is impossible, then subsequent tries add the constraint $z \neq D\,(\overline{D})$.

If all tries prove impossible, then the fault is redundant. If for all POs no solution was found after 3 round-robin turns each (or, more exactly 2+4+8 = 14 seconds), then the fault is aborted.

Results for the ATG system under strategy number 1 are depicted in Table 3.9 (obtained on the same platform as the experiences without ITBS of Table 3.8).

Table 3.9. ATG results with ITBS over sensitised POs

	Red.	Ab.	#	Time	T/TG
c432	1	3	89	53	0.6
c499	8	0	94	84	0.8
c880	0	0	87	13	0.1
c1355	8	0	149	108	0.7
c1908	6	2	171	249	1.4
c2670	105	39	177	1958	6.1
c3540	129	6	266	1041	2.6
c5315	59	0	179	1149	4.8
c6288	34	60	24	10960	92.9
c7552	131	12	335	7230	15.1

This table shows that ITBS clearly outperforms the cardinality constraint approach in this case. These results are already competitive with those of Simonis, which although obtained in a slower computer, took advantage of its RISC architecture and a more efficient Prolog compiler, together with specialised ATG heuristics. The number of aborted faults explains the results obtained with circuit *c6288* where, without specialised heuristics, 8 seconds were not enough for any of the candidate POs (which in this circuit were generally several).

ITBS parameters can be adapted to each circuit since larger circuits usually require more computation time in TG. The "tougher" *c6288* circuit could benefit from increasing the maximum limit of 8 seconds, and a lot of time would be saved if the rounds of 2 and even 4 seconds were simply abolished since it is rare to find solutions inside those times. This is particularly relevant when there is a large number k of disjuncts, which is also usually the case with this circuit due do its higher average fanout (see Table 2.2).

Since a more constrained subproblem may be easier to solve due to higher propagation and reduced search space, we can use strategy number 2 (specify PO) to extend further the disjunction of having a sensitised PO into the disjunction where a PO may assume either value D or $\overline{D}$. This value specification may double the number of disjuncts, each corresponding to a more constrained subproblem. Since we specify the PO value, we may also completely specify a sensitised path all the way down to the S-buffer, and then try to justify it as before. We tested ITBS over this extended disjunction to obtain the results of Table 3.10.

Results are largely similar in both cases. Extending the disjunction is sometimes better, because an easier subproblem is found (which was the rationale of ITBS in the first place). However, it can also be worse, because the number of subproblems that are explored increases, and so does total computation time.

Table 3.10. ATG results with ITBS over specified sensitised POs

	Red.	Ab.	#	Time	T/TG
c432	1	3	102	128	1.2
c499	8	0	110	26	0.2
c880	0	0	95	17	0.2
c1355	8	0	180	141	0.8
c1908	6	2	173	179	1.0
c2670	105	11	229	1079	3.1
c3540	129	6	296	6327	14.7
c5315	59	0	191	4097	16.4
c6288	34	79	27	28662	204.7
c7552	131	11	338	9099	19.0

There is a trade-off between the number of disjuncts, which further constrain subproblems (in the limit it would just be *generate-and-test*), and *least commitment* where no choices whatsoever are made and all decisions delayed.

3.6.1 Conclusion

This section has shown that ITBS is in general effective for this kind of problems. With ITBS we were able to use a better variable ordering heuristic and compensate for the fact that no value ordering heuristic was used. This was possible due to the nature of ITBS. If a wrong choice is made, ITBS is not stuck at it, but rather abandons it after a limited amount of time.

ITBS will thus be used in the following chapters for related problems.

3.7 Summary

This chapter presented a basic problem of digital circuits and showed the adequateness of multi-valued logics and constraint programming to model and solve it, together with a generalisation to MSFs. In addition, a general search strategy, iterative time-bounded search, was proposed to solve disjunctive goals and its effectiveness was demonstrated for test generation in combinational circuits by obtaining significantly better results on a set of benchmark circuits.

The next chapter addresses a more complex problem: circuit diagnosis. In particular, the problem of correctly diagnosing a faulty circuit by differentiating two possible sets of faults is modelled and solved using constraints over an 8-valued logic that extends the 4-valued logic by encoding one extra type of dependency (on the other set of faults).

Chapter 4

Differential Diagnosis

In this chapter we address the issue of locating faults at the logic level in a 'black box' circuit. In general, diagnosis of a system presenting an unintended behaviour consists of identifying the faulty components, and their faulty states. More specifically, the problem of **differential diagnosis** of faulty gates in a VLSI circuit, where the only observable findings are its input/output bits, consists of finding tests that allow the differentiation of two alternative diagnoses. Being able to locate the precise cause of an incorrect response at the production stage may improve the manufacturing process thus reducing future costs. Also, when a system in regular use eventually becomes faulty, it may be important to determine what is the exact diagnosis. For example, if a circuit is not accessible and cannot be replaced (e.g. in space), it might be important to assess which functions are affected by the faulty gate(s). Faulty circuits, although with a more limited capacity, may often be still useful if its limitations are understood.

To model these problems we developed an eight-valued logic that describes the dependency of the findings on the competing diagnoses. We realised later that the basis of such logic was already described in [Abramovici *et al.* 1990] albeit incompletely, so we formally define and discuss the rationale for a complete eight-valued logic. Additionally, we implemented a constraint solver to handle this eight-valued logic efficiently, namely to obtain differentiating tests that allow the elimination of one of the alternative diagnoses. We discuss the limitations of the techniques currently available to handle disjunctive constraints, and use the previously described iterative time-bounded search (ITBS) method to overcome them.

The chapter is organised as follows. After a brief introduction in section 4.1 and a discussion of approaches in section 4.2, section 4.3 describes differential diagnosis and test patterns in more detail. Section 4.4 then presents the 8-valued logic showing its use in modelling the generation of differential test patterns in combinational circuits before an alternative logic is commented in section 4.5. Section 4.6 describes a constraint solver to handle directly the 8-valued logic. Section 4.7 describes specialised benchmarks and section 4.8 differentiation algorithms. Section 4.9 then presents experimental results before conclusions are summarised in the last section.

4.1 Introduction

When the output of some system does not correspond to its expected behaviour for a given input, one is faced with the problem of diagnosis, which corresponds to locating (i.e. identifying) the fault(s) responsible for the incorrect output. In a faulty digital circuit, many single or multiple faulty gates may explain the observed findings. We will refer to each such possible set of faults as a diagnostic set. Diagnosis is a task that may be performed in a variety of forms but in abstract it involves two sub-tasks: firstly, one has to obtain a number of hypotheses that explain the observed findings; secondly, one has to differentiate the alternative diagnoses. Given its purpose of differentiating hypotheses, this sub-task is usually referred to as *differential diagnosis*.

This chapter will focus on differentiating the available candidates, i.e. in the sub-task of differential diagnosis. Notice that the techniques (discussed in the previous chapter) that generate test patterns for verifying whether one specific gate is faulty do not necessarily identify patterns that discriminate between alternative diagnostic sets. Such patterns were generated assuming that either the gate to be tested is faulty or not, and all other gates are normal. In this context, for a certain test pattern, inspection of the output of the circuit unveils the state of the gate under test. Unfortunately, these assumptions are not sufficient for differential diagnosis, as one is interested in differentiating between alternative hypotheses. It is very common that, given two alternative faulty gates, each of them has a large number of test patterns, but only a few of them, if any, do differentiate the faults. In general, a test pattern detects a number of possible diagnostic sets (sets of faulty gates). Hence, two diagnostic sets may have a large number of test patterns in common and yet be indistinguishable (i.e. for the same inputs they both entail the same output, albeit not the correct one). For example, in a small available [ISCAS 1985] circuit (*voter_flat*) with 12 input bits and 4 output bits, we found that two alternative faults had each 121 test patterns, but they were in fact indistinguishable, i.e. no input pattern could unveil (by only measuring its input and output) which of the faults might be actually present in the circuit. Although, in this small example all these patterns could be generated in a few seconds, larger circuits with thousands of gates and tens to hundreds of inputs and outputs could simply not be tested.

Test patterns that allow to discard hypothetical faults contribute to improve the *diagnostic resolution*, i.e. the degree of accuracy to which faults can be located. The *maximal fault resolution* of a system is defined by the partition of all the possible faults into distinct sets of functionally equivalent faults. No external testing experiment can distinguish among functionally equivalent faults, and determining whether two arbitrary faults are functionally equivalent is an NP-complete problem [Goundan 1978].

In the general Test Generation (TG) problem, the goal is to find a test pattern that logically entails a different output from a circuit where specific faults are present. The problem that is the topic of this chapter, **Differential Test Generation** (DTG), aims at generating patterns for a circuit that would entail different outputs for different sets of faulty gates.

A differential test pattern is then an input vector of the circuit that induces different Boolean values in some output bit for the two different diagnostic sets of arbitrary size. Such pattern allows the elimination of a diagnostic candidate requiring no expensive generate-and-test procedure.

4.2 Diagnosis Approaches

Given its declarative nature, expressive power and efficient execution, Constraint Logic Programming may be used in various diagnostic (sub-)tasks, namely in the context of digital circuits. To obtain a candidate hypothesis, it is sufficient to represent each digital component, usually a gate, as a disjunction of constraints, each representing a possible behaviour of the component. By providing the observed input and output, the CLP system solves the constraints, and inspection of the solutions unveils which of the components are faulty.

Handling disjunctions usually leads to lengthy computations due to its combinatorial nature. Moreover, one is usually interested in hypotheses that are minimal, given some definition of minimality. In the context of digital circuits one is usually interested in hypotheses that contain a minimum number of faulty gates. Minimal solutions, which may be enforced in a CLP program as an optimisation problem, are a natural feature of Hierarchical Constraint Logic Programming (HCLP). In this extension, a constraint is either mandatory or not, and the solutions reported for a problem are minimal, i.e. they discard minimal sets of non-mandatory constraints (see [Borning *et al.* 1989] for details).

In an implementation of an HCLP system for Boolean constraints [Menezes and Barahona 1996] minimal diagnostic sets can be efficiently obtained for the benchmark ISCAS digital circuits. Given some input, if the value of some output bit(s) is different from normal, then several diagnoses explain the abnormal findings. Generation of candidate diagnoses can be declaratively modelled in an HCLP(B) system. In such a system, Boolean circuit gates can be expressed as defeasible (non-mandatory) constraints that can be relaxed if necessary. To obtain the faulty gates in the HCLP system, when the behaviour of the circuit is different from what is expected, one has simply to provide the inputs and outputs of the circuit, as well as its model (a set of defeasible constraints) and obtain the (minimal) sets of constraints that are relaxed [Menezes and Barahona 1996]. Assuming that the only two faulty modes for a gate are the usual stuck-at-0 and stuck-at-1 (i.e. the output of a gate is, respectively 0 or 1, regardless of its input), then each relaxed constraint may be interpreted as faulty in the state opposed to the normal output. This can be illustrated with the full adder circuit of Figure 4.1:

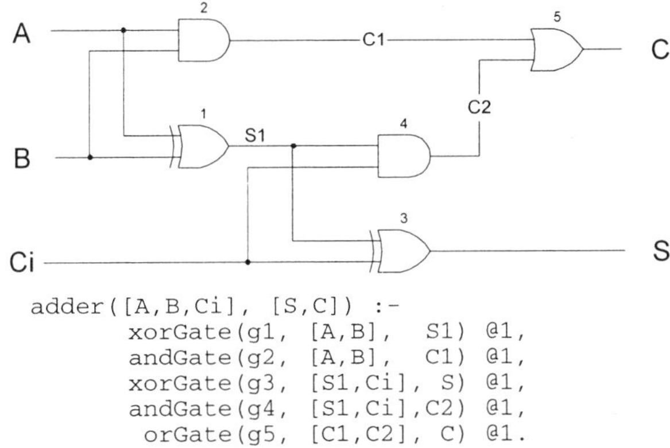


```
adder([A,B,Ci], [S,C]) :-
        xorGate(g1,  [A,B],   S1) @1,
        andGate(g2,  [A,B],   C1) @1,
        xorGate(g3,  [S1,Ci],  S) @1,
        andGate(g4,  [S1,Ci], C2) @1,
         orGate(g5,  [C1,C2],  C) @1.
```

Figure 4.1. Full adder circuit and its specification

The $@L$ operator simply specifies that a constraint is of level L (mandatory if $L=0$); in this case, the gates are all defeasible with the same likelihood ($@1$). Additionally, each gate has an extra argument that uniquely identifies the constraint. In this circuit we may have the following query/answers:

```
Query ?- adder([0,0,0], [1,0]).

Answer 1:  relaxed g1([0,0], 1)
Answer 2:  relaxed g3([0,0], 1)
```

The answer thus shows (by backtracking) that there are two minimal hypotheses (single faults), which occur by malfunctioning of gates 1 or 3. By inspecting the output of the relaxed gates one may immediately conclude that the single faults are either gate *g1* stuck-at-1 or gate *g3* stuck-at-1 (abbreviated to *g1*/1 and *g3*/1, respectively). Thus, the problem of differential diagnosis, i.e. differentiating the alternative hypotheses, becomes relevant.

4.3 Differential Diagnosis and Test Patterns

In some cases, some differentiation of candidate diagnostic sets is obtained during the process of generating them, by directing the search of candidates so as to generate the most interesting candidates first [Damásio *et al.* 1995]. This preference may be expressed in terms of some likelihood measurement of the candidates (e.g. their probability [Peng and Reggia 1991]). For the precise differentiation of diagnoses, these techniques are not adequate. We are not interested in choosing more likely or preferred diagnosis, but rather to know, for certain, whether candidates can be eliminated (for example, in order to still use the non-affected functions of a VLSI circuit).

In other cases, the differentiation requires the execution of extra tests in order to obtain more information that allows the elimination of hypotheses (see [Console and Frederic 1994] and [Dressler 1997] for a brief survey on a number of different techniques that may be used). This is the approach followed here, where the differential test patterns form the extra tests required.

These test patterns may be used to eliminate candidate hypotheses (if the observed output is not the same as predicted by the (faulty) model, then the fault may be discarded). In the context of single fault diagnosis, in the example above, input [0,0,0] is a test pattern for both faults *g1*/1 and *g3*/1, but it does not provide information to differentiate the two. This can be done, in this case, with pattern [0,0,1], which for a normal circuit outputs [1,0], for a circuit with fault *g1*/1 outputs [0,1], and for a circuit with fault *g3*/1 outputs [1,0]. Hence, execution of this extra test allows discarding one of the hypotheses.

Finding a test pattern that does not yield the same output for two faults (i.e. a differential test) may be hard to find, namely if a generate-and-test approach is used, since, in the worst case, all the test patterns of both faults have to be generated and tested.

This is particularly inefficient when the faults are indistinguishable, as is the case with faults *g2*/1, *g4*/1 and *g5*/1. In this case they all havè only a few common test patterns ([0,0,0], [0,1,0], [1,0,0] and [0,0,1]) and so it is not hard to generate and check it. In other cases with larger circuits, the situation is much worse and the naïve generate-and-test approach becomes naturally infeasible.

To differentiate two sets of faults in a digital circuit, i.e. to check which is actually present, it is necessary to confirm whether, for other input vectors, the real circuit behaves as having either of the faults. (The models of the two possible faulty circuits are easily obtained by replacing the normal gates in the original model with the faulty ones.) In the worst case, one may have to consider all input vector configurations. In general, a generate-and-test approach is not feasible for large circuits because the input vectors that might have to be tested are exponential on the number of PIs. Some of these vectors mask the effects of the faults, which are not propagated to the output. Hence, we only need to consider those input vectors that propagate a fault.

Current techniques deal with differential diagnosis by trying to generate input vectors that cause different outputs in two circuits (among the normal and the alternative abnormal circuits). In [Hartanto *et al.* 1997] one tries to detect one fault assuming the circuit has the other; in [Gruning *et al.* 1991] one tries to detect one fault without detecting the other; finally, in [Pomeranz and Reddy 1998] one tries to detect both and after undetect one of them. Nevertheless, the complexity of the diagnosis increases significantly with the extra circuitry involved. Modelling such problems into a SAT solver poses similar problems regarding the multiplication of circuits [Manquinho and Silva 2000], and the associated combinatorics.

The technique we developed and present in this chapter, can be regarded as an extension of those explored in the previous chapter (incidentally, it solves the problem of the *voter_flat* circuit, referred in section 4.1, in a few milliseconds). We extended the notion of dependencies further to differentiate two alternative models, and introduced an 8-valued logic whose values not only denote dependency on faulty gates, but also discriminate the dependencies between two sets of faulty gates [Azevedo and Barahona 1998], i.e. multiple faults. A differential test pattern is then an input vector that logically entails one such dependency-discriminating value in a PO.

Despite being inspired in that 8-valued logic, a previous system we developed [Azevedo and Barahona 1998] was implemented with a "traditional" 0/1 Boolean solver, i.e. the 6 extra values were assigned to digital signals by means of logic disjunctions, only remaining constraints on the Boolean domain.

4.4 The 8-valued Logic

The purpose of our 8-valued logic is to allow tracing the dependency of a truth-value on the alternative diagnostic sets, or simply diagnoses, that are to be differentiated. It is important to notice the context in which this differentiation takes place.

To start with, model N of a circuit is assumed to have no faults. When an output O is observed which is not consistent with the input I and circuit model N, such output can be explained by alternative diagnoses that justify the anomalous finding. Let us consider two such diagnoses, ϕ_1 and ϕ_2, each consisting of some faulty gates. Given input I, the two models obtained by replacing the normal gates of N by the faulty gates ϕ_1 or ϕ_2, will be referred to as F_1 and F_2, respectively, and both logically entail output O. Since both models logically entail an output different from that entailed by the normal model N, I is an input test pattern for both ϕ_1 and ϕ_2. However, I does not enable the differentiation between ϕ_1 and ϕ_2, since both corresponding models yield the same output from input I. We are thus concerned with finding a differential test pattern D that yields two different outputs, O_1 and O_2, for the two faulty models F_1 and F_2.

Outputs O_1 and O_2 are different if at least one of the output bits is different, hence such bit depends on either ϕ_1 or ϕ_2, but not on both. The logic presented in this section captures this notion of dependency by means of its 8 values. Each value is denoted either by Boolean X or by a pair $<p\text{-}X>$: p ranges over the set $\{m, d_1 \text{ and } d_2\}$ and denotes the dependency on the alternative diagnoses; X ranges over the usual 0/1 truth-values and is the "physical" truth-value that would be observed if the circuit were normal (model N). The meaning of the 8 values is thus the following:

X the truth value is independent from faults ϕ_1 and ϕ_2. These faults have no influence on the physical truth-value which is always X.

$d_1\text{-}X$ the truth value depends on faults ϕ_1 but not on ϕ_2. In model F_1, where ϕ_1 occurs, the physical truth-value is the complement of X; otherwise (in both models N and F_2), it is X.

$d_2\text{-}X$ similar to the previous, with ϕ_1 and ϕ_2 swapping roles.

$m\text{-}X$ the truth value depends on both diagnoses ϕ_1 and ϕ_2. In both models F_1 and F_2, the physical truth-value is the complement of X, obtained in model N.

With such definitions a differential test pattern is an input vector of the circuit that entails an output bit with a d-value (i.e. a value $d_S\text{-}X$, where $S \in \{1,2\}$ and $X \in \{0,1\}$), whose physical value depends on one and only one of the two sets of faults. As such, similarly to simple TG, an error coming from ϕ_1 (ϕ_2) must be propagated to a PO thus forming a sensitised path for ϕ_1 (ϕ_2). We call such a path, a **differential path** for ϕ_1 with respect to ϕ_2. Clearly, if all the gates in a sensitised path for ϕ_1 are not sensitised for ϕ_2, then the path is differential.

At first thoughts, one could be led to think that all the gates of a sensitised path for ϕ_1 should not be sensitised for ϕ_2 in order to guarantee their differentiation. The intuition is that if some gate in a sensitised path for ϕ_1 were also sensitised for ϕ_2, then it would propagate to the subsequent gates in the path, not only fault(s) ϕ_1 but also fault(s) ϕ_2, making such path useless to differentiate the two diagnoses.

However, imposing that *all* the gates in the path are independent from ϕ_2 is an unnecessarily strong condition, as can be seen in the circuit of Figure 4.2 where $\phi_1 = \{x/0\}$ and $\phi_2 = \{y/0\}$ are the candidate diagnoses. Under test $t=11$, *nand*-gate n is dependent on both $x/0$ and $y/0$, as its output becomes 1 when either x or y are stuck-at-0. Nevertheless, output z is the opposite of x's S-buffer*, regardless of whether x is stuck-at-0 (in which case y is normal: Figure 4.2 (b)) or is normal (and y is normal or stuck-at-0: Figure 4.2 (a) and (c)). Despite in the sensitised path for diagnosis $\{x/0\}$ under t one of its gates, n, is not independent from the alternative diagnosis $\{y/0\}$, one may still consider such path as a differential path for $\{x/0\}$ with respect to $\{y/0\}$ under test t: in fact, one may conclude that if the output bit is physically 0 then x is necessarily normal, whereas if the bit is 1 then y is necessarily normal.

---

* S-buffers were discussed in the previous chapter: Test Patterns

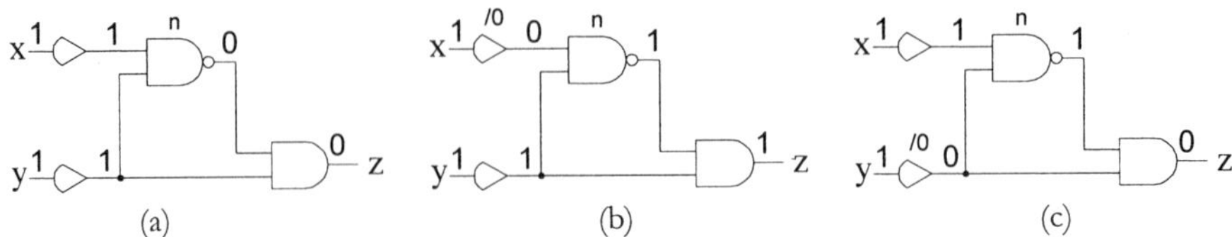

Figure 4.2. Differential test for $x/0$ and $y/0$; a) normal circuit N; b) F_1 with $x/0$; c) F_2 with $y/0$

Let us analyse the problem of encoding, in general, the truth-value Z of some signal line in this 8-valued logic. Table 4.1 shows the 8 possible cases.

Table 4.1. Truth-value of a signal with a normal model and two different diagnoses

N	0	0	0	0	1	1	1	1
F_1	0	0	1	1	0	0	1	1
F_2	0	1	0	1	0	1	0	1
Z	0	d_2-0	d_1-0	m-0	m-1	d_1-1	d_2-1	1

In the first column of the table, the physical truth-value is the same (0) in all cases, i.e. when either one or none of the diagnoses ϕ_1 and ϕ_2 are considered. In the second column, the physical truth-value is 0 in the initial model N and in F_1, but 1 in F_2. It thus depends on faults ϕ_2 (but not on ϕ_1) and is encoded as d_2-0. In the fourth column, the physical truth-value is 0 in the normal model N but 1 in both F_1 and F_2. It is thus dependent on both diagnostic sets ϕ_1 and ϕ_2 and is encoded as m-0.

The example of Figure 4.2 may therefore be compactly represented as in Figure 4.3.

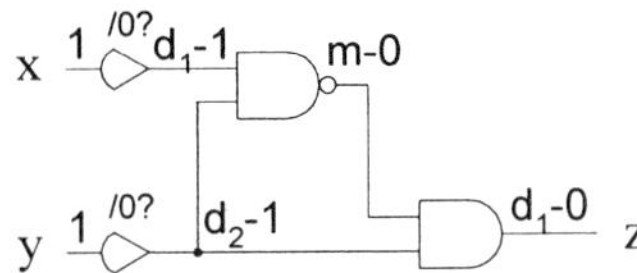

Figure 4.3. Circuit with 8-valued logic for $x/0$ or $y/0$

4.4.1 Boolean operations

A similar analysis can be performed to define the semantics of the usual Boolean operations. For example, the semantics of exclusive disjunction (*xor*) for the 8-valued logic is shown in Table 4.2.

Table 4.2. XOR truth table in 8-valued logic

xor	m-0	d_2-0	d_1-0	0	1	d_1-1	d_2-1	m-1
m-0	0	d_1-0	d_2-0	m-0	m-1	d_2-1	d_1-1	1
d_2-0	d_1-0	0	m-0	d_2-0	d_2-1	m-1	1	d_1-1
d_1-0	d_2-0	m-0	0	d_1-0	d_1-1	1	m-1	d_2-1
0	m-0	d_2-0	d_1-0	0	1	d_1-1	d_2-1	m-1
1	m-1	d_2-1	d_1-1	1	0	d_1-0	d_2-0	m-0
d_1-1	d_2-1	m-1	1	d_1-1	d_1-0	0	m-0	d_2-0
d_2-1	d_1-1	1	m-1	d_2-1	d_2-0	m-0	0	d_1-0
m-1	1	d_1-1	d_2-1	m-1	m-0	d_2-0	d_1-0	0

As expected, *xor*-ing independent values (0/1) produces independent values according to the usual *xor* truth tables (centre of Table 4.2). *Xor*-ing a d_1 signal (truth value) and a d_2 signal results in an *m* signal reflecting the dependency on both ϕ_1 and ϕ_2. Due to the nature of the exclusive disjunction, when both inputs depend on the same diagnostic set, the output of the gate does not depend on it, and is thus an independent signal (cf. the diagonals of Table 4.2). Perhaps more interestingly, *xor*-ing an *m* signal (dependent on both diagnostic sets ϕ_1 and ϕ_2) with a signal that only depends on one set makes the output solely dependent on the other set.

This latter case is explained in Table 4.3 below. Columns N, F_1 and F_2 represent the cases to consider. The first two lines represent the physical truth-values which are *xor*-ed: input signal *m-1*, takes the physical truth value 1 in model N, but 0 in both F_1 and F_2 (*m*-signals depend on both faulty components); input signal d_1-0, takes the physical truth-value 0 in models N and F_2, but 1 in F_1 (it depends only on ϕ_1 as a d_1 signal). The physical truth-value of the output is obtained by *xor*-ing the physical truth-values of the inputs. Since the physical output is 1 in both models N and F_1, but 0 in F_2 it is encoded as d_2-*1*.

Table 4.3. Physical output of an *xor*-gate with inputs *m-1* and d_1-*0*

	N	F_1	F_2
m-1	1	0	0
d$_1$-0	0	1	0
m-1 $\oplus$ d$_1$-0	1	1	0

Similar reasoning was used to define the semantics for the extended 8-valued logic of all the operations implemented by the usual digital gates.

4.4.2 Modelling Alternative Diagnostic Theories in Digital Circuits

We now show how to model a faulty digital circuit for which there are two alternative diagnoses available. As in the previous chapters, faulty gates will be modelled by means of S-buffers. According to the previous notation, model N corresponds to the circuit where all the S-buffers are functioning correctly. In models F_i, (*i* in {1,2}), some gates are stuck and the output of the corresponding S-buffers have a fixed physical truth-value. S-buffers have specific truth tables since they can be faulty. An S-buffer can be so because it belongs to set of faults ϕ_1 or ϕ_2 or to both. The fault can be stuck-at-0 or stuck-at-1 in any one of the sets. It is possible for an S-buffer to appear in both sets being stuck-at-0 in one and stuck-at-1 in the other (of course, this will only happen with diagnostic sets consisting of 2 or more faults). There are thus 8 possibilities of dependency of an S-buffer to consider. Table 4.4 below shows the resulting model for all S-buffers.

Table 4.4. Truth table for the 8 different types of S-buffer

Dependency \ Input	m-0	d$_2$-0	d$_1$-0	0	1	d$_1$-1	d$_2$-1	m-1
$F_1/0, F_2/0$	0	0	0	0	m-1	m-1	m-1	m-1
$F_2/0$	d$_1$-0	0	d$_1$-0	0	d$_2$-1	m-1	d$_2$-1	m-1
$F_1/0$	d$_2$-0	d$_2$-0	0	0	d$_1$-1	d$_1$-1	m-1	m-1
$F_1/0, F_2/1$	*d$_2$-0*	d$_2$-0	d$_2$-0	d$_2$-0	d$_1$-1	d$_1$-1	d$_1$-1	d$_1$-1
$F_1/1, F_2/0$	d$_1$-0	d$_1$-0	d$_1$-0	d$_1$-0	d$_2$-1	d$_2$-1	d$_2$-1	d$_2$-1
$F_1/1$	m-0	m-0	d$_1$-0	d$_1$-0	1	1	d$_2$-1	d$_2$-1
$F_2/1$	m-0	d$_2$-0	m-0	d$_2$-0	1	d$_1$-1	1	d$_1$-1
$F_1/1, F_2/1$	m-0	m-0	m-0	m-0	1	1	1	1

The table entries are explained with the example shown in italic and shaded. The fourth line corresponds to an S-buffer that is stuck-at-0 in F_1 and stuck-at-1 in F_2, and the first entry in that line corresponds to its output, given an input signal *m-0*. Being an *m* signal, the input of the gate denotes that it depends on both faults ϕ_1 and ϕ_2 (more precisely, since the circuits under consideration are purely combinational, the signal depends on other faulty gates in diagnostic sets ϕ_1 and ϕ_2). This case is illustrated in Table 4.5.

Table 4.5. Output of an S-buffer (stuck-at-0 in F_1 and stuck-at-1 in F_2) for input *m-0*

	S-buffer ($F_1/0, F_2/1$)	
	m-0	output
N	0	0
F_1	1	0
F_2	1	1

The first column corresponds to the physical truth-value of the buffer input, which is 0 in model N and 1 in both F_1 and F_2. In case of model N (no faults) the buffer behaves correctly and outputs the same value of the input. In F_1 the buffer is stuck-at-0 and thus outputs 0. Finally, in F_2 the buffer is stuck-at-1 and thus outputs 1. Inspection of the output shows that it is 0 in both models N and F_1, but 1 in F_2; it is thus encoded as d_2-0.

4.5 A 4-valued logic for Differentiation

The 8-valued-logic allows the modelling of two faulty circuits, in conjunction with the normal one, by expressing signal dependencies on them. This extends and doubles the 4-valued logic to handle one single diagnostic set, making it possible to differentiate two such sets in addition to simple TG. Note that TG for a diagnostic set ϕ is simply modelled with the 8-valued logic by the differentiation of ϕ with the empty diagnostic set {}. Then, only 4 values $\{0, 1, d_1$-$0, d_1$-$1\}$ are possible, and the goal of having a *d*-signal at the output ensures that the test pattern detects ϕ.

Similarly, if one only wants to differentiate two alternative diagnoses, a simpler logic encoding just the two corresponding faulty circuits could be used. In this situation, the normal circuit can be dropped since all that is necessary is to keep track of signal differences between the two faulty circuits. With one less circuit, pairs of values from the 8-valued logic will have the same meaning since there is no need to consider the 0- and 1- values of the normal circuit. Thus, this logic can be collapsed into a 4-valued logic for differentiation purposes, as shown in Table 4.6 where each of the 4 values $\{00, 11, 01, 10\}$ represents 2 values of the 8-valued logic (column *8V*).

Table 4.6. Condensation of 8-valued logic into a 4-valued logic for differentiation

F_1	F_2	8V
0	0	0, *m*-1
1	1	1, *m*-0
0	1	d_1-1, d_2-0
1	0	d_1-0, d_2-1

For instance, value 00 represents 8-valued logic values 0 and *m*-1, since these are those encoding digital signals with a 0-value in both faulty circuits (with faults ϕ_1 and ϕ_2). 8-valued logic value 0 represents constant 0 in the 3 circuits (including the normal one), and *m*-1 is used for signals that are normally 1 but take value 0 in the two faulty circuits (the signals are dependent on both diagnoses). Hence, 0 and *m*-1 signals take value 0 in both F_1 and F_2. Similar cases occur for other logic values with the extraction of just the values for F_1 and F_2.

Logic operations over this logic can be trivially performed by separately considering the 2 circuits, to yield tables such as in Figure 4.4.

NOT			AND	00	11	01	10		OR	00	11	01	10
00	11		**00**	00	00	00	00		**00**	00	11	01	10
11	00		**11**	00	11	01	10		**11**	11	11	11	11
01	10		**01**	00	01	01	00		**01**	01	11	01	11
10	01		**10**	00	10	00	10		**10**	10	11	11	10

Figure 4.4. Logic operations over a 4-valued logic for diagnosis

Therefore, two circuits (either normal and faulty circuits, or two faulty circuits) can be modelled by a 4-valued logic, thus enabling TG and DTG to be handled with the same complexity.

Although it can lead to some performance loss on differentiation applications, we focus mainly on the implementation of the 8-valued logic since it is more general (the 4-valued logic may be collapsed from it and is equivalent to the one presented for TG in Chapter 3) and will be useful to better understand other logics presented in the next chapters.

4.6 A Constraint Solver for the 8-Valued Logic

In the 8-valued logic, 6 extra values corresponding to the d and m signals are added to the usual Boolean 0/1 values. In a previous approach [Azevedo and Barahona 1998], despite using an 8-valued logic, we addressed the problem of differential diagnosis of digital circuits with a 0/1 Boolean solver (the extra 6 values were propagated to the output, by analogy with the demons approach taken in [Simonis 1989] but with choice points to cope with the different possibilities) implemented in SICStus Prolog [SICStus 1995].

In order to avoid choice points in the selection of differential paths (a typical generate and test approach suffering from well known efficiency problems discussed in section 3.4.1.2), one should use constraints over the 8 values, in a constraint programming philosophy. We thus incorporated the 6 special signals with the Boolean values into a single 8-valued domain. To implement such a finite domain constraint solver, it was necessary to adapt the truth tables of the usual Boolean operators and the S-buffers for the particular encoding of the 8-valued logic, and also to specify a constraint propagation strategy.

Arc consistency (see section 1.2) is computationally too demanding in general, especially if more than just binary constraints are handled, as already shown in the 4-valued logic for TG discussed in Chapter 3. With larger domains, the inadequacy of such kind of consistency tends to increase. Nevertheless, a reasonable amount of propagation should be ensured, if only without much effort, and maintaining node consistency is often not sufficient for adequate propagation. Additionally, a simple and easy to maintain implementation was sought, for general purpose problems involving 3 circuits, taking full advantage of the underlying SICStus CLP(FD) solver which has the ability to accept user-defined constraints. The way we achieved these objectives is described in the next section.

In our new solver, the basic gates modelled as user-defined constraints are '*not*', '*and*' and '*xor*' in addition to the trivial '*buffer*'. Other types of gates such as '*or*', '*nor*' and '*nand*' are modelled in terms of the basic ones, and can have an arbitrary number of inputs, since our '*and*' constraint definition handles this case.

All user-defined constraints that implement this 8-valued logic are activated by guards, normally fired at instantiation of one of its variables. In addition to these higher-level constraints, cardinality constraints were used, as well as constructive disjunction [le Provost and Wallace 1993] for some relations between 2 variables. By so doing, the constraint solver imposes a form of consistency stronger than node consistency, but somewhat weaker than arc consistency (too costly to maintain). Whenever two variables alone remain in a constraint, arc-consistency is usually maintained; otherwise, in general, propagation is triggered by instantiation of a variable.

We now examine in more detail the solver implementation.

4.6.1 Domain Representation

Since we were using SICStus that provides a CLP(*FD*) solver library [Carlsson *et al.* 1997] over integers, we had first to encode the 8-valued domain $D_8 = \{0,1,d_1\text{-}0,d_1\text{-}1,d_2\text{-}0,d_2\text{-}1,m\text{-}0,m\text{-}1\}$ into an integer domain, in order to allow (built-in) arithmetic constraints between all the values.

A simple encoding of D_8 using integers 0 to 7, keeping the 0-1 Boolean values followed by the other signals, would make it hard to express even simple constraints as *'not'* and would force us to handle unnecessary disjunctions which would make computation harder. E.g. if the 8 values $\{0,1,d_1\text{-}0,d_1\text{-}1,d_2\text{-}0,d_2\text{-}1,m\text{-}0,m\text{-}1\}$ were encoded respectively as $\{0,1,2,3,4,5,6,7\}$ then their negation (*'not'* operation) values would be $\{1,0,3,2,5,4,7,6\}$, and the relation $Y = not(X)$ may be expressed by the (long) disjunction:

$$(X = 0 \wedge Y = 1) \vee (X = 1 \wedge Y = 0) \vee (X = 2 \wedge Y = 3) \vee (X = 3 \wedge Y = 2) \vee \ldots\ldots \vee (X = 7 \wedge Y = 6)$$

Since 'negated' values are 'neighbours', we could represent the relation in a simpler form, stating that the 'distance' between X and Y is 1. However, wrong neighbours (e.g. 1 and 2, 3 and 4, 5 and 6) must be avoided. A possibility of avoiding such cases could be by means of explicit "primitive" constraints as in

$$(X - Y = 1 \vee Y - X = 1) \wedge X + Y \neq 3 \wedge X + Y \neq 7 \wedge X + Y \neq 11$$

This is clearly a very heavy constraint to process efficiently, and better alternatives were explored.

Ideally, any logic operation should be transformed in a simple arithmetic constraint. In the particular case of binary constraints in the form of an equation involving just sum or subtraction operations, SICStus ensures arc-consistency with little computational effort. Nevertheless, for such equations, other CLP solvers can still efficiently apply only bounded arc-consistency (section 1.3.1), in which the constraint is triggered only by changes in the bounds of domain variables (and also only bounds become updated). In any case, the code is kept simple and a sufficiently efficient processing underneath is assured.

We could thus think of a domain such as $\{-4,-3,-2,-1,1,2,3,4\}$ for, respectively, the 8 values $\{m\text{-}0,d_2\text{-}0,d_1\text{-}0,0,1,d_1\text{-}1,d_2\text{-}1, m\text{-}1\}$, which would allow the simple constraint $Y = -X$ to model a *not*-gate with input X and output Y.

Let us now examine the respective *xor* table (depicted in Table 4.7).

Table 4.7. XOR for encoded domain $\{-4...-1, 1..4\}$

xor	m-0 -4	d₂-0 -3	d₁-0 -2	0 -1	1 1	d₁-1 2	d₂-1 3	m-1 4
-4	-1	-2	-3	-4	4	3	2	1
-3	-2	-1	-4	-3	3	4	1	2
-2	-3	-4	-1	-2	2	1	4	3
-1	-4	-3	-2	-1	1	2	3	4
1	4	3	2	1	-1	-2	-3	-4
2	3	4	1	2	-2	-1	-4	-3
3	2	1	4	3	-3	-4	-1	-2
4	1	2	3	4	-4	-3	-2	-1

With such 'symmetric' encoding, we also obtain simple relations for the *xor* operation (see Table 4.8 considering each possible *xor* output value). Each relation between two variables corresponds to the constraint we have to impose when another variable becomes instantiated. Remember that our goal is to find simple constraints for pairs of variables, which can possibly be easily implemented with arc-consistency (or approximated) with the underlying CLP tool.

Table 4.8. Output of *xor*-gate implies relation between arguments

X⊕Y	-4	-3	-2	-1	1	2	3	4
X,Y	|X+Y|=5	|X-Y|=2	|X-Y|=1	X=Y	X=-Y	|X+Y|=1	|X+Y|=2	|X-Y|=5

Unfortunately, the relations of Table 4.8 hold only in an implication sense — it is not always an equivalence relation. In other words, for a given *xor* output, the corresponding relation between arguments is only a necessary condition for that output to hold — it is not always a sufficient condition.

For example, with output d_1-0 (encoded as –2), it is indeed necessary that the 'distance' (absolute difference) between arguments X and Y is 1 (possible pairs: {-4,-3}, {-2,-1}, {1,2}, {3,4}). However, with this encoding, there are other possible pairs that still verify this condition but where the exclusive disjunction does not yield the same output value. E.g., with arguments $(X,Y)=(-3,-2)$ verifying the same distance (1), the output is different: $Z= X\oplus Y = -4$ (i.e. *m*-0). Hence, either additional constraints would have to be imposed to disallow such pairs or the original *xor* constraint should remain in store after propagation.

To automatically discard such undesirable extra pairs of values we tried to add 'holes' in the encoded domain. Then, values that respect some distance relation to argument X would either entail the given *xor* output value or fall into a domain hole, which corresponds to an impossible assignment. Hence, to approximate at least bounded arc-consistency with a small effort, as we will next see, the domain of codes $CD_8 = \{-6,-5,-3,-2,2,3,5,6\}$ was adopted with the interpretation shown in Table 4.9.

Table 4.9. Adopted 8-valued domain encoding

value	m-0	d_2-0	d_1-0	0	1	d_1-1	d_2-1	m-1
code	-6	-5	-3	-2	2	3	5	6

The order of values is preserved (with respect to the previous encoding) but the values are further separated. In the following sections we discuss the different gate constraints with this domain.

4.6.2 Not-Gates

With this domain, the negation of a variable X (in a *'not'* gate, as in Table 4.10) is the symmetrical arithmetic value (see Table 4.11). Hence the output of a gate receiving as input some signal X {-6, -5, -3, -2, 2, 3, 5, 6} is simply replaced by $-X$. The constraint propagation mechanism used maintains arc-consistency and is standard in SICStus for finite domains (domains maintained as indexical constraints, as first adopted in constraint logic programming in the language CLP(*FD*) [Diaz and Codognet 1993]).

Table 4.10. 8-valued logic table for NOT

X	m-0	d_2-0	d_1-0	0	1	d_1-1	d_2-1	m-1
not(X)	m-1	d_2-1	d_1-1	1	0	d_1-0	d_2-0	m-0

Table 4.11. 8-valued logic encoded table for NOT

X	-6	-5	-3	-2	2	3	5	6
not(X)	6	5	3	2	-2	-3	-5	-6

4.6.3 Xor-Gates

The '*xor*' logic operation of Table 4.2 may be summarised as in Table 4.12 where, for each of the 8 possible outputs, we have a list of 8 possible input pairs, noticing that '*xor*' is commutative.

Table 4.12. XOR possible input pairs for each output

$X \oplus Y$	m-0	d_2-0	d_1-0	0	1	d_1-1	d_2-1	m-1
X,Y	{0, m-0}	{0, d_2-0}	{0, d_1-0}	{S, S} :	{0, 1}	{0, d_1-1}	{0, d_2-1}	{0, m-1}
	{1, m-1}	{1, d_2-1}	{1, d_1-1}	$S \in D_8$	{d_1-0,d_1-1}	{1, d_1-0}	{1, d_2-0}	{1, m-0}
	{d_1-0,d_2-0}	{d_1-0,m-0}	{d_2-0,m-0}		{d_2-0,d_2-1}	{d_2-0,m-1}	{d_1-0,m-1}	{d_1-0,d_2-1}
	{d_1-1,d_2-1}	{d_1-1,m-1}	{d_2-1,m-1}		{m-0,m-1}	{d_2-1,m-0}	{d_1-1,m-0}	{d_1-1,d_2-0}

Converting into our encoded domain we obtain Table 4.13 below.

Table 4.13. XOR possible encoded input pairs for each output

$X \oplus Y$	-6	-5	-3	-2	2	3	5	6
X,Y	{-2, -6}	{-2, -5}	{-2, -3}	{S, S} :	{-2, 2}	{-2, 3}	{-2, 5}	{-2, 6}
	{2, 6}	{2, 5}	{2, 3}	$S \in CD_8$	{-3, 3}	{2, -3}	{2, -5}	{2, -6}
	{-3, -5}	{-3, -6}	{-5, -6}		{-5, 5}	{-5, 6}	{-3, 6}	{-3, 5}
	{3, 5}	{3, 6}	{5, 6}		{-6, 6}	{5, -6}	{3, -6}	{3, -5}

An '*xor*' is now also easily expressed by Table 4.14 where each output specifies an arithmetic constraint over the inputs.

Table 4.14. Relation between output of an *xor*-gate and its arguments

$X \oplus Y$	-6	-5	-3	-2	2	3	5	6
X,Y	$\|X+Y\|=8$	$\|X-Y\|=3$	$\|X-Y\|=1$	$X=Y$	$X=-Y$	$\|X+Y\|=1$	$\|X+Y\|=3$	$\|X-Y\|=8$

Notice that $X \oplus Y = Z \Leftrightarrow X \oplus Z = Y \Leftrightarrow Z \oplus Y = X$. Hence, as in the case of a *not*-gate, propagation handles inputs and outputs alike. The absolute value expressions, such as $|X+Y|=8$, are modelled as (constructive) disjunctions ($X+Y=8 \vee X+Y= -8$).

Now, let us assume that initially X, Y and Z had all CD_8 as domain. Suppose that, at some point, Y is instantiated to value 3 (i.e. d_1-1). We then replace the *xor* constraint by ($X+Z=1 \vee X+Z= -1$). X and Z must still keep CD_8 as their domain since all values are yet possible (due to the nature of *xor*), but as soon as, for instance, value 6 (i.e. *m*-1) is removed from the domain of X, we can remove value 1-6 = -5 (i.e. d_2-0) from the domain of Z. Hence, due to the constraint implementation, arc-consistency is ensured by a simple constructive disjunction. Notice that such simple disjunction would not be enough with the simpler encoding (e.g. {–4..-1,1..4}) previously discussed. The removal of value 4 (i.e. *m*-1) from the domain of X, after Y is instantiated to value 2 (i.e. d_1-1) would not impose the removal of value 1-4 = -3 (i.e. d_2-0) from the domain of Z since this value (-3) would still have support in value 2 (i.e. Boolean 1) of X, allowed by the disjunctive constraint $|X+Z|=1$, since one also has -1-2 = -3 (i.e. $|4+(-3)|=1$ and $|2+(-3)|=1$).

4.6.4 Normal And-Gates

Table 4.15 presents the semantics of the *and*-operation ($A = X \wedge Y$) in the 8-valued logic.

Table 4.15. AND (8-valued) logic table

$A = X \wedge Y$	m-0	d_2-0	d_1-0	0	1	d_1-1	d_2-1	m-1
m-0	m-0	d_2-0	d_1-0	0	m-0	d_2-0	d_1-0	0
d_2-0	d_2-0	d_2-0	0	0	d_2-0	d_2-0	0	0
d_1-0	d_1-0	0	d_1-0	0	d_1-0	0	d_1-0	0
0	0	0	0	0	0	0	0	0
1	m-0	d_2-0	d_1-0	0	1	d_1-1	d_2-1	m-1
d_1-1	d_2-0	d_2-0	0	0	d_1-1	d_1-1	m-1	m-1
d_2-1	d_1-0	0	d_1-0	0	d_2-1	m-1	d_2-1	m-1
m-1	0	0	0	0	m-1	m-1	m-1	m-1

The propagation mechanism that we implemented delays propagation until one of the arguments is ground. Table 4.16 shows, for each ground value of X, what can be inferred on input Y and output A. The possible pairs (A,Y) that are the solutions of each set of inferences (a line in the table) are exactly those that can be taken from examining Table 4.15, and thus the original 'and' constraint can be removed from the store and replaced by the new ones.

Table 4.16. Constraints when input X of A= X∧Y is known

X	Constraints on {A, Y} for A = X∧Y
m-0	A ∈ {m-0,d₂-0,d₁-0,0}, A=Y ∨ (Y,A) ∈ {(1,m-0), (d₁-1,d₂-0), (d₂-1,d₁-0), (m-1,0)}
d₂-0	A ∈ {d₂-0,0}, (Y ∈ {m-0, d₂-0,1,d₁-1}, A=d₂-0) ∨ (Y ∈ {d₁-0,0,d₂-1,m-1}, A=0)
d₁-0	A ∈ {d₁-0,0}, (Y ∈ {m-0, d₁-0,1,d₂-1}, A=d₁-0) ∨ (Y ∈ {d₂-0,0,d₁-1,m-1}, A=0)
0	A= 0
1	A=Y
d₁-1	A ∈ {d₂-0,0,d₁-1,m-1}, (Y ∈ {m-0, d₂-0}, A=d₂-0) ∨ (Y ∈ {d₁-0,0}, A=0) ∨ (Y ∈ {1,d₁-1}, A=d₁-1) ∨ (Y ∈ {d₂-1,m-1},A=m-1)
d₂-1	A ∈ {d₁-0,0,d₂-1,m-1}, (Y ∈ {m-0, d₁-0}, A=d₁-0) ∨ (Y ∈ {d₂-0,0}, A=0) ∨ (Y ∈ {1,d₂-1}, A=d₂-1) ∨ (Y ∈ {d₁-1,m-1},A=m-1)
m-1	A ∈ {0,m-1}, (Y ∈ {m-0, d₂-0,d₁-0,0}, A=0) ∨ (Y ∈ {1,d₁-1,d₂-1,m-1}, A=m-1)

Table 4.17 includes both previous tables using our codes and showing the arithmetic constraints that we post on each case. For example, if X takes value 5 (code for d_2-1), the domain of A is reduced to {-3,-2,5,6} and, additionally, the constraint $A=Y \vee A-Y=3$ is posted to replace the original 'and' constraint. (Similarly to the *xor* constraint, it can be verified that this disjunction captures exactly all possible pairs {A,Y}. As can be seen in Table 4.16, when $X= d_2$-1, for each possible value of A, there are only two possible values of Y. Thanks to our encoding, the difference A-Y between A and Y can only be 0 or 3.)

Table 4.17. 'And' encoded logic table and constraints

A = X∧Y	-6	-5	-3	-2	2	3	5	6	Constraints on {A,Y}
-6	-6	-5	-3	-2	-6	-5	-3	-2	A<0, A=Y ∨ Y-A=8
-5	-5	-5	-2	-2	-5	-5	-2	-2	A ∈ {-5,-2}, Y-A ∈ {-1,0,7,8}
-3	-3	-2	-3	-2	-3	-2	-3	-2	A ∈ {-3,-2}, Y-A ∈ {-3,0,5,8}
-2	-2	-2	-2	-2	-2	-2	-2	-2	A= -2
2	-6	-5	-3	-2	2	3	5	6	A=Y
3	-5	-5	-2	-2	3	3	6	6	A ∈ {-5,-2,3,6}, A=Y ∨ A-Y=1
5	-3	-2	-3	-2	5	6	5	6	A ∈ {-3,-2,5,6}, A=Y ∨ A-Y=3
6	-2	-2	-2	-2	6	6	6	6	A ∈ {-2,6}, (Y<0,A= -2) ∨ (Y>0,A=6)

The whole propagation mechanism (including constructive disjunction to handle the above disjunctions) was implemented with the user-defined primitives available on SICStus.

Of course, similar behaviour is specified for the case where Y is ground (X replacing Y in the table). The case where the output is ground has to be treated separately. Table 4.18 shows the constraints that are posted and replace the original one, when the value of output A becomes instantiated (the case $A = -2$ is treated differently, being handled as a mere propagation rule, i.e. the original 'and' constraint is kept in the constraint store).

Table 4.18. Constraints to post when output of *and*-gate becomes instantiated

'And' output signal	output code *A*	Constraints on inputs *X* and *Y*
m-0	-6	X,Y in {-6,2}, X+Y ≠ 4
d_2-0	-5	X,Y in {-6,-5,2,3}, X+Y ≤ -2, X+Y ≠ -12, X+Y ≠ -4
d_1-0	-3	X,Y in {-6,-3,2,5}, X+Y ≤ 2, X+Y ≥ -9, X+Y ≠ -4
0	-2	X+Y ≤ 4, X+Y ≥ -8, and(X,Y)= -2
1	2	X=Y=2
d_1-1	3	X,Y in {2,3}, X+Y ≠ 4
d_2-1	5	X,Y in {2,5}, X+Y ≠ 4
m-1	6	X+Y ≥ 8, X+Y ≠ 10

And-gates with *k* inputs (*k*>2) could of course be handled as *k-1*, 2-input *and*-gates. E.g. $A = X \wedge Y \wedge Z$ handled as $A = and(X, and(Y, Z))$). However this would not allow efficient propagation in some cases, namely when *A* becomes ground (it would directly affect only *X* and the output of the other *and*-gate which could remain uninstantiated, thus unable to constrain inputs *Y* and *Z*). Hence, each gate with more than 2 inputs is implemented as a global constraint [Beldiceanu 1990]. In such case, all inputs (*X,Y,Z*) can be directly constrained, but only the '*in*' component of the constraints in Table 4.18 is posted, i.e. domain of the input variables is only narrowed, without adding any more constraints to the store. Furthermore, there are cases where it is possible to constrain inputs to have at least one particular value. For instance, if output is *m-1* and *and*-ing the already ground inputs yields d_1-1 (d_2-1), then one of the other inputs must be d_2-1 (d_1-1). Simple analysis to the truth table confirms it. Such disjunction can be imposed by means of a cardinality constraint [Van Hentenryck and Deville 1991] and is implemented in SICStus using its *count/4* constraint.

4.6.5 S-Buffers

The truth tables defining S-buffers, shown in Table 4.4, are rewritten in Table 4.19, with the CD_8 encoding, relating the S-buffer output *B* to its input *I* (heading row). It also shows the constraints used to define these buffers (according to each case of dependency).

Arithmetic constraints are derived in a similar way to normal gates. Since the dependency type of an S-buffer is known at 'compile-time' (the two diagnoses, ϕ_1 and ϕ_2, are given to our DTG system), only two variables may remain (input and output), which allows us to fully replace S-buffers constraints by simple arithmetic constraints (disjunctions are handled by user-defined constraints) as soon as the former are posted, similarly to *not*-gates.

This concludes the description of gate constraints. All these user-defined constraints are activated by guards, usually fired at instantiation of one of its variables. In addition to these higher-level constraints, cardinality constraints as well as constructive disjunction and conjunction were used as described in the previous tables.

All such constraint propagation is far more "complete" than node-consistency (where propagation is only triggered when all but one of the variables in a constraint are ground). It is also more complete than bounded arc-consistency, which is usually maintained by the underlying CLP tool on general arithmetic constraints. Nevertheless, it will not enforce full arc-consistency, as the trade-off between the pruning that it could achieve and the cost of implementing it is not worth it.

For example, when one of the variables in a binary '*xor*' constraint takes value 0 (the guard checks this condition) the constraint is simply rewritten as an equality of the other two variables. As another example, if one variable takes value *m-1*, there are eight possible pairs of values for the other two variables, and the '*xor*' constraint is thus rewritten into the constructive disjunction of these cases. In a final example, if the output of an *and*-gate takes value 0, the '*and*' constraint is fired and may be replaced by the cardinality constraint imposing that at least one of the inputs be 0 (in fact, there are some combinations of d_i and m signals that also yield value 0 when 'anded'; the '*and*' constraint checks this possibility).

Table 4.19. Truth table and constraints for the 8 different types of S-buffer

Dependency	m-0 -6	d₂-0 -5	d₁-0 -3	0 -2	1 2	d₁-1 3	d₂-1 5	m-1 6	Constraints
$F_1/0, F_2/0$	-2	-2	-2	-2	6	6	6	6	$B \in \{-2,6\}$, $(I<0, B = -2) \lor (I>0, B = 6)$
$F_2/0$	-3	-2	-3	-2	5	6	5	6	$B \in \{-3,-2,5,6\}$, $B-I \in \{0,3\}$
$F_1/0$	-5	-5	-2	-2	3	3	6	6	$B \in \{-5,-2,3,6\}$, $B \geq I, B \leq I+1$
$F_1/0, F_2/1$	-5	-5	-5	-5	3	3	3	3	$B \in \{-5,3\}$, $(I<0, B = -5) \lor (I>0, B = 3)$
$F_1/1, F_2/0$	-3	-3	-3	-3	5	5	5	5	$B \in \{-3,5\}$, $(I<0, B = -3) \lor (I>0, B = 5)$
$F_1/1$	-6	-6	-3	-3	2	2	5	5	$B \in \{-6,-3,2,5\}$, $I \geq B, I \leq B+1$
$F_2/1$	-6	-5	-6	-5	2	3	2	3	$B \in \{-6,-5,2,3\}$, $I-B \in \{0,3\}$
$F_1/1, F_2/1$	-6	-6	-6	-6	2	2	2	2	$B \in \{-6,2\}$, $(I<0, B = -6) \lor (I>0, B = 2)$

4.6.6 Heuristics to Find Differential Patterns

As explained, a differential test pattern must logically entail a d_i signal in the digital circuit output with all gate constraints posted. An output bit must therefore have a value in the set $\{-5,-3,3,5\}$ (otherwise the two sets of faults were indistinguishable) so that this bit depends exclusively either on set ϕ_1 (values -3 and 3) or on set ϕ_2 (-5 and 5). As in the simple TG problem (Chapter 3), this is the typical disjunctive constraint suitable for the iterative time-bounded search (ITBS) technique (described in section 3.6) that we propose for this kind of problems and applied for finding differential test patterns.

We tried three different scenarios to confirm the best heuristic. In the first, we used the least commitment approach, implemented with a cardinality operator on the output bits that included a d value in their domains after setting up the circuit. To speed up this execution, we labelled first the input bits that lead to the possible S-buffers (i.e. in their transitive fanin), and then the inputs that lead to the possible output bits. This is the Cardinality (#) heuristic.

The second scenario adopted the ITBS strategy with a heuristic that is similar to the previous one, but which is applied only after committing to one of the output bits. This is the ITBS-Bit strategy. The (potential) advantage now is that by committing to some output bit, more propagation is possible in principle and fewer input bits are relevant, which thus decreases the search space.

The third scenario, ITBS-Path, also used the ITBS technique but with a different heuristic (Path) that is obtained after selecting a definite d-signal to an output bit. Since this d-signal depends either on faulty components F_1 (values d_1-0 and d_1-1) or on faulty components F_2 (d_2-0 and d_2-1), it is then mandatory that the corresponding S-buffers produce the d signal thus activating a fault. This is imposed with a cardinality constraint, implemented with the built-in constraint *element/3*. To reach the output bit, there must be a differential test path, i.e. a path from an S-buffer to that bit [Azevedo and Barahona 1998]. With this knowledge, we first try to find that path by labelling the variables starting in the d-signal "backwards" to an S-buffer in a way that they remained always dependent on F_1 or F_2. Since the FD solver is not complete, we then try to justify the chosen path by labelling the input bits that are relevant to it, i.e. that allow the d signal to reach the output bit. We did it starting at the S-buffer (the path start gate) transitive fanin PIs, which are then labelled first. We then followed along the chosen differential test path. For each gate in the path, we would similarly find the input bits of the circuit that are relevant for this gate, and label them in the second place. This process finishes when reaching the final path bit (a circuit output).

ITBS-Path is very similar to the discussed TG heuristic of completely specifying a sensitised path. Note that ITBS-Bit does not stick to any path at all, but only to its final bit.

For any heuristic, it is sufficient to label the Boolean circuit input bits and have all variables instantiated to assure a solution. Some bits, however, have nothing to do with the differential path (i.e. are outside of its transitive fanin) and could lead to irrelevant backtracking. Hence, we adopted the strategy of labelling first the circuit inputs relevant to this path.

All this labelling may be done with the following heuristics: the most constrained variables are chosen first; the values are chosen with arbitrary domain enumeration. Finally, to have a completely specified test pattern, we label the remaining inputs.

4.7 Benchmarks

Before discussing results, we first describe the set of experiments that adopted the ISCAS circuits (section 2.4) to obtain such results with our system. Unfortunately, there are, to our knowledge, no standard benchmarks for the problem of differentiating faulty diagnoses. Therefore we had to create, for each circuit, specific benchmarks consisting of a set of competing diagnoses to differentiate between [Azevedo and Barahona 1999]. We describe this process first.

4.7.1 Generating a Benchmark

In general it is easy to differentiate "unrelated" faulty gates. The problem becomes more difficult, if the competing diagnostic sets are known to have produced the same faulty output for some input. Hence, we have assumed that the most difficult situations arise when the two sets of faults to differentiate result from diagnosis of a malfunctioning circuit.

These diagnostic sets typically contain the same number n of faults, since one is usually interested in minimal (with respect to cardinality) diagnosis (usually $n=1$, corresponding to SSF). To produce more "realistic" benchmarks for sets of n faults on each circuit c, we proceeded in the following way:

1- generate a random input In as a list of 0-1 values (considered equiprobable) for the PIs of c;
2- with the model of c with no faults, generate its "normal" output Out as a list of 0-1 values for the POs of c;
3- randomly invert (equiprobably) n bits of Out to yield "faulty" output $FOut$;
4- with some diagnostic tool (see section 4.2), generate the set S_n of all minimal diagnoses that explain $FOut$ given input In

Our diagnostic tool (using HCLP(B) system of [Menezes and Barahona 1996]) examined directly the ISCAS circuit model, and assumed that each gate (including PIs) could be either stuck-at-0 or stuck-at-1. No special fault set for the circuit was considered, since emphasis was on differentiation.

Example. As an example, Table 4.20 shows 2 inputs randomly generated for circuit *c432* and the corresponding normal outputs:

Table 4.20. Random inputs generated for circuit c432, and corresponding outputs

Input	Normal Output
$I_a=$ 1100100111001101011011111110001111111	1111101
$I_b=$ 1100100011110111001011100100011110111	1101010

With these inputs, benchmarks were created as described above. We show the first tries in Table 4.21.

Table 4.21. Benchmarks attempts for c432

Input	Faulty Output	Fault Type / Inverted POs	Diagnoses	Benchmark
I_a	1111100	single / 7	$S_1 = \{\{432\text{gat}/0\}\}$	No benchmark
I_a	1110101	single / 4	$S_1 = \{\{380\text{gat}/0\}, \{415\text{gat}/1\}, \{416\text{gat}/1\}, \{421\text{gat}/0\}\}$	b432_a_1
I_a	1011001	double / 2 and 5	$S_2 = \{\{187\text{gat}/0,430\text{gat}/0\}, \{270\text{gat}/1,430\text{gat}/0\}, \{329\text{gat}/0,430\text{gat}/0\}, \{37\text{gat}/1,105\text{gat}/0\}, \{43\text{gat}/0,105\text{gat}/0\}, \{47\text{gat}/1,430\text{gat}/0\}\}$	b432_a_2
I_b	1101000	single / 6	$S_1 = \{\{419\text{gat}/0\}, \{428\text{gat}/1\}, \{431\text{gat}/0\}\}$	b432_b_1a
I_b	1111010	single / 3	$S_1 = \{\{370\text{gat}/1\}, \{92\text{gat}/0\}\}$	b432_b_1b

The first experiment with input I_a with the 7th bit of its normal output inverted, generated no benchmark since there was only one possible diagnosis for that faulty behaviour, namely gate '432gat' stuck-at-0. The second attempt with input I_a resulted in a set of 4 diagnostic sets (each with one faulty gate). The third, resulted in a set of 6 diagnostic sets (each with 2 faulty gates). Similar experiments were done with input I_b.

4.7.2 Set of Benchmarks Used

Each of our benchmarks thus consists of a set of all minimal diagnoses (the result of the diagnostic process) that explain some faulty behaviour of a circuit. We concentrated only on single and double fault diagnoses. For each of the ISCAS circuits (from *c432* to *c7552*) we generated two different inputs, and for each input, we generated two single fault outputs and one double fault output, thus producing a total of 6 benchmarks per circuit (4 sets S_1 plus 2 sets S_2). For the larger ISCAS circuits, starting with *c1908*, only single fault benchmarks were produced. In total, we generated 51 benchmarks, 43 regarding diagnostic sets with single faults and 8 with double faults.

The rationale behind this was to get a representative but not too big set of benchmarks, especially for single faults, since they all are basically a particular application of the same general principle.

All benchmarks have names of the form *b<Circuit>_<InputId>_<FaultsNumber>* possibly followed by an identifier to remove ambiguities.

4.8 Differentiating Multiple Diagnoses

In this section, we present two simple algorithms that systematically generate differential test patterns from a set of candidate diagnoses. The first, more exhaustive, algorithm (for benchmark purposes) partitions the initial set of candidate diagnoses into classes of indistinguishable diagnoses without the need for testing a physical circuit. A second algorithm eliminates incorrect diagnoses by applying generated differential tests in the physical circuit in order to obtain a minimal subset of indistinguishable hypotheses that explain the circuit's faulty behaviour. We briefly describe the algorithms and, in the next section, present the experimental results obtained in the set partitioning together with proposals for algorithm improvements.

Clearly, the indistinguishable relation (let us denote it by $d_1 =_d d_2$) between two diagnoses, d_1 and d_2, is reflexive ($d =_d d$), symmetric ($d_1 =_d d_2 \Leftrightarrow d_2 =_d d_1$) and transitive ($d_1 =_d d_2 \land d_2 =_d d_3 \Rightarrow d_1 =_d d_3$). As such, it defines classes of indistinguishable diagnoses. This justifies that the algorithms presented in this section (indeed any algorithm based on the same assumptions that only the input and output bits are observable) will terminate with sets of candidate diagnoses that cannot be further reduced.

For benchmark purposes, to make the testing of the differentiation problem exhaustive we decided to compute the set of all the equivalence classes of indistinguishable diagnoses, in order to achieve the maximal fault resolution. This is done with the *Classes* algorithm shown below (pseudo-code in Figure 4.5), which simply stores each of the diagnostic sets in the corresponding class by trying to generate differential tests to conclude whether two diagnoses are indistinguishable.

```
Procedure Classes (In: Si, Out: Cs);
   Cs ← ∅                    % no classes yet.
   while Si ≠ ∅ do
      Si ← Si \ {D₁}         % obtain a diagnosis D₁.
      Cs' ← Cs               % classes to check in loop.
      Found ← ∅              % no equivalence class found so far.
      while Cs' ≠ ∅ and Found = ∅ do
         Cs' ← Cs' \ {C}     % obtain a class C.
         select D₂ from C    % pick any of its diagnoses.
         if not Diff_Pattern(D₁,D₂,I) then Found ← {C}
               % equivalence class found.
      end while
      if Found = {C}        % an equivalence class C for D₁ was found.
         then Cs ← Cs \ C ∪ {C ∪ {D₁}}    % update class C in Cs.
         else Cs ← Cs ∪ {{D₁}}            % add new class.
   end while
end Procedure
```

Figure 4.5. Algorithm to partition a set into classes of indistinguishable diagnoses

Procedure *Classes* gets as input a set *Si*, of candidate hypotheses (diagnostic sets) and, trivially, picks one at a time to check if it belongs to any of the current computed classes. Checking if a diagnosis D_1 belongs to some class C is performed by picking any diagnosis D_2 of C and verifying if D_1 and D_2 are indistinguishable (by failing to find a differentiating test I with predicate *Diff_Pattern(D₁,D₂,I)*). If no equivalent class is found, then a new one with current diagnosis D_1 is added to the current set of classes. The final set of classes (each a set of diagnoses) is returned as parameter *Cs*.

In principle, diagnoses of benchmarks such as those described in the previous section are harder to differentiate as they are tightly related (in particular, recall that a single input *In* is used to generate *Si*, hence *In* is a test pattern for all the elements of *Si*).

The critical part of the algorithm is the computation of differential patterns, which can be performed from $\#Si - 1$ times (in the best case, i.e. a single class) to $\#Si * (\#Si - 1) / 2$ times (in the worst case, i.e. no equivalent diagnoses, which corresponds to $\#Si$ classes).

For the example of the full-adder of Figure 4.1, with input 000 and output 10 the possible minimal diagnostic sets were *{g1/1}* and *{g3/1}*. Given input $Si = \{\{g1/1\}, \{g3/1\}\}$, the *Classes* algorithm just has to generate a differential test pattern I for this pair of faults (with *Diff_Pattern({g1/1},{g3/1},I)*) to conclude they are differentiable and obtain two classes, i.e. $Cs = \{\{\{g1/1\}\},\{\{g3/1\}\}\}$. In the same circuit, with input 000 and output 01, the possible faults are *g2/1*, *g4/1* and *g5/1*, which are indistinguishable. Given $Si = \{\{g2/1\}, \{g4/1\}, \{g5/1\}\}$, we thus obtain $Cs = \{\{\{g2/1\}, \{g4/1\}, \{g5/1\}\}\}$ by twice attempting and failing to generate a differential test pattern.

Clearly, in the worst case, DTG is exponential on the number of input bits. However, our experience shows that in practice, with our tool, even for circuits with hundreds of input bits, differential patterns may be found effectively, as we will see in the next section.

In 'real life', with the real circuit to test, there is no need to obtain all equivalence classes since only one matters: the single class that explains the circuit behaviour given any input. The algorithm for discriminating the possible diagnoses from an initial set of hypotheses is shown in Figure 4.6. The algorithm makes pairwise comparisons of all the candidate diagnoses, and outputs those that were not eliminated (they all belong to the same class). For each pair of candidates, it tries to find a differential test pattern for them. If it fails, then the diagnoses are in the same class of indistinguishable diagnoses; otherwise, it applies this pattern in order to eliminate one of the candidates.

```
Procedure Differentiate (In: Si, Out: So);
    Si ← Si \ {D}        % remove a candidate D from the set.
    C ← {D}              % initialise the candidate class with it.
    while Si ≠ ∅ do
        select D₁ from C    % pick a candidate diagnosis D₁.
        Si ← Si \ {D₂}      % obtain an alternative candidate D₂.
        if Diff_Pattern(D₁, D₂, I) then   % differentiable by I.
            if Real_Output(I) ≠ Simulated_Output(D₁,I) then %not F₁
                C ← {D₂}     % make {D₂} the new current class.
            end if
        else C ← C ∪ {D₂}   % add D₂ to the current set of
        end if               % indistinguishable candidates.
    end while
    So ← C
end Procedure
```

Figure 4.6. Algorithm to obtain a set of indistinguishable diagnoses

Procedure *Differentiate* gets as input a set *Si*, of candidate hypotheses (diagnostic sets), and systematically makes pairwise comparisons of these possible diagnoses. At any time, a set C maintains the set of (indistinguishable) candidates that are the current solution. If a differential test pattern I, between a member D_1 of C and an alternative diagnosis D_2, is found (with predicate *Diff_Pattern(D₁,D₂,I)*) it is applied to the circuit. Then, if the output in the real circuit *Real_Output(I)* is different from the output *Simulated_Output(D₁,I)* that would be obtained with the diagnosis D_1 of the current solution set, then candidate D_1, as well as all the other candidates in set C are discarded and $\{D_2\}$ becomes the current set of candidates. Otherwise candidate D_2 is simply discarded. If, on the contrary, a differential test between D_1 and D_2 was not found, then they are indistinguishable and D_2 is added to current set C. Eventually, all initial candidates are tested and set C, returned as the answer, contains all the indistinguishable diagnoses that explain the faulty behaviour of the circuit for any input.

It is easy to prove that the algorithm is correct. Its complexity is linear on the number of initial hypotheses, since all of them are tested once in the internal loop. (This is in contrast with the *Classes* algorithm, where for each initial hypothesis, all current computed classes are potentially considered with a differentiation, thus achieving quadratic complexity.)

Returning to the example of the full-adder of Figure 4.1, with $Si = \{\{g1/1\}, \{g3/1\}\}$, the *Differentiate* algorithm just has to generate a differential test pattern I for this pair of faults (with $Diff\_Pattern(\{g1/1\}, \{g3/1\}, I)$) to conclude which one is normal and which one is actually faulty (by comparing the real and simulated outputs given input I).

When the possible faults are $g2/1$, $g4/1$ and $g5/1$, which are indistinguishable, input $Si = \{\{g2/1\}, \{g4/1\}, \{g5/1\}\}$ is given to the *Differentiate* algorithm to obtain $So = \{\{g2/1\}, \{g4/1\}, \{g5/1\}\}$ by 2 failed DTG attempts.

4.9 Experimental Results

We now report the performance of our system on the set of differentiation benchmarks described in section 4.7.

4.9.1 Choosing the Heuristic

In section 4.6.6 we described three heuristics, namely the *Cardinality* (#) heuristic (using a cardinality constraint over the POs), *ITBS-Bit* (ITBS committing to a PO) and *ITBS-Path* (ITBS committing to a d-signal and choosing a path). To check which heuristic was the best, we tested these different strategies in the problem of finding differential test patterns to verify whether the ITBS technique was indeed appropriate for it.

For each differentiaton problem, the number of variables is the number of circuit signals, whereas the number of constraints is dependent on the type of the circuit gates, since gates (constraints) such as '*or*' and '*nand*' are modelled as combinations of basic constraints '*and*' and '*not*'. While $nand(\{I_1, I_2\})$ is modelled with 2 constraints as $not(and(\{I_1, I_2\}))$, a gate representing $or(\{I_1, I_2, I_3\})$ is modelled with 5 constraints as $not(and(\{not(I_1), not(I_2), not(I_3)\}))$. A *nor*-gate is, in addition, modelled as the negation of an *or*-gate. Hence, the number of constraints in DTG problems for circuits having, for example, a large number of *nor*-gates, such as *c6288*, largely exceeds the number of original circuit gates. More precisely, for such *nor*-gates, the number of constraints is $N_{nor} * (Av_{fanin}+3)$, where N_{nor} is the number of *nor*-gates in the circuit and Av_{fanin} its average fanin, since each input bit contributes with one constraint, whereas the output contributes with 2 (*not*-gates), and the gate itself with 1 (the basic '*and*'). Notice that '*xor*' is also implemented as a basic gate.

Table 4.22 below shows some of the experimental results obtained with our solver, a system implemented in SICStus over Linux on a PentiumIII/500 (reported times are all in seconds). We chose a small yet representative set of differentiation problems for single and double faults over a small circuit (*c432*) and a large circuit (*c6288*), which are usually the hardest circuits in their size class.

The two sets of faults ϕ_1 and ϕ_2 that we want to differentiate consist of one or more faults in the form *gate/stuck-at-value*; N_o is the number of output bits that remain with d-values in their domains after all the gate constraints have been posted and propagated; N_i is the number of input bits that must be labelled to guarantee the solution; TT is the total time needed to obtain the solution with ITBS, *Lim* being the time limit when a solution was found (the initial time limit of ITBS was 5 seconds); *bit* indicates on which of the possible output bits the solution was found; and d indicates the number of the successful d-value choice. Non-available values represent aborted executions.

Table 4.22. Differentiation results with the different heuristics

| | | | # | | ITBS | | | | | | | |
| | | | | | Bit | | | | Path | | | |
circuit	$\Phi1$	$\Phi2$	No	Time	Ni	bit	Lim	TT	Ni	d	Lim	TT	Ni
432	47/1,430/0	270/1,430/0	3	n/a	36	3	5	10.2	36	2	5	5.2	36
432	223/0,338/1	223/0,319/0	4	396.4	36	2	5	5.3	36	2	5	5.2	36
432	223/0,430/1	223/0,338/1	4	0.2	36	2	5	5.3	36	2	5	5.3	36
432	223/0,386/1	223/0,319/0	4	373.0	36	2	5	5.2	36	2	5	5.2	36
432	37/1,105/0	270/1,430/0	7	n/a	36	1	5	0.3	19	1	5	0.3	18
432	329/0,430/0	270/1,430/0	5	0.2	36	1	5	0.2	27	1	5	0.2	27
6288	3486gat/0	2434gat/1	23	4.2	32	1	5	4.1	20	3	5	14.0	20
6288	5348gat/1	5163gat/1	7	4.8	32	1	5	4.8	32	2	10	74.8	32
6288	5461gat/0	4808gat/1	7	4.8	32	1	5	4.8	32	5	5	24.9	32
6288	6285gat/0	5727gat/1	2	5.3	32	1	5	5.8	32	1	5	5.7	32
6288	1173gat/0	1128gat/0	17	4.8	32	1	5	4.6	32	1	5	4.8	32
6288	1546gat/1	1343gat/1	23	n/a	32	n/a	n/a	n/a	n/a	1	5	4.0	20

4.9.2 Discussion

The cardinality operator to deal with disjunctive constraints is often quite effective. However, in a significant number of cases the results are quite disappointing. In the smaller circuit, *c432*, the first test shown could not be solved in any acceptable time limit (tens of hours). In fact, by not committing to any of the 3 output bits where a *d*-signal occurs, no significant pruning of the search space (2^{36}) was achieved. This is not the case with the ITBS method where such commitment is enough to propagate enough information to prune the search space. Of course, the most effective choice was not always considered first. But here the time-bounded nature of the ITBS method avoids a strong commitment to the wrong choice. For example, ITBS with the *Bit* heuristic easily finds a solution in the 3rd choice, being interrupted after 5 seconds of fruitless search in the previous 2 choices. Similarly, ITBS with the *Path* heuristic finds a solution with the second choice of a *d*-signal.

It is not simple to analyse how the pruning takes place. In general, it is due to constraint propagation, as the number of bits to label is the same in all heuristics. There are some exceptions though. In one of the experiments with the *c432* circuit, committing to some output bit immediately reduces the number of bits to label from 36 to 19 (ITBS-Bit) or 18 (ITBS-Path) with the corresponding efficiency improvement.

Of course, no heuristic is always guaranteed to succeed. Even when there is a commitment to some output bit (and hence a stronger heuristic (*Bit*) compared to the case with no commitment (#)), this stronger heuristic is not guaranteed to find a solution (e.g. last line in Table 4.22). However, using the other more specialised heuristic (*Path*) one was always able to find a solution (sometimes only at the cost of being diverted from the wrong choices by the time limits of ITBS).

This confirms the importance of the ITBS technique as an alternative to the cardinality operator. The former allows the use of more powerful heuristics in the labelling phase of problem solving. The risks associated to the wrong choices are softened by the time-bounded commitment to them.

4.9.3 Complete Results

To test the performance of our system with *ITBS-Path*, we tried to obtain, for each differentiation benchmark, a partition of the set of diagnostic sets, into classes of indistinguishable diagnostic sets with the *Classes* algorithm (Figure 4.5). The results for each benchmark are shown in Table 4.23, where N is the initial number of possible diagnoses; *NC* is the resulting number of distinct classes; *ND* is the number of differentiation tests needed to reach the final partition; *TT* is the total time (in seconds) spent for that; Ta is the average differentiation time and Te is the expected time to find the final minimal diagnostic class for a real circuit, which is given by $Te = (N\text{-}1) * Ta$, since $N\text{-}1$ pairwise comparisons (differentiations) have to be performed as in the *Differentiate* algorithm (Figure 4.6).

Table 4.23. Complete results for differentiation benchmarks

	N	NC	ND	TT	Ta	Te
b432_a_1	4	1	3	0.35	0.12	0.35
b432_a_2	6	5	11	7.71	0.70	3.50
b432_b_1a	3	2	2	0.37	0.19	0.37
b432_b_1b	2	2	1	0.22	0.22	0.22
b432_b_2	4	4	6	16.40	2.73	8.20
b432_c_1	2	2	1	0.21	0.21	0.21
b499_a_1a	3	3	3	7.37	2.46	4.91
b499_a_1b	6	6	15	11.23	0.75	3.74
b499_a_2	9	9	36	82.86	2.30	18.41
b499_b_1a	3	3	3	0.58	0.19	0.39
b499_b_1b	3	3	3	6.15	2.05	4.10
b499_b_2	48	46	1037	1727.82	1.67	78.31
b880_a_1a	4	2	3	0.94	0.31	0.94
b880_a_1b	6	3	6	2.51	0.42	2.09
b880_a_2	33	5	56	22.38	0.40	12.79
b880_b_1a	3	2	2	0.80	0.40	0.80
b880_b_1b	2	1	1	0.29	0.29	0.29
b880_b_2	96	12	380	163.38	0.43	40.85
b1355_a_1a	6	4	10	7.32	0.73	3.66
b1355_a_1b	5	3	7	435.42	62.20	248.81
b1355_a_2	36	12	102	7300.80	71.58	2505.18
b1355_b_1a	27	16	137	8066.32	58.88	1530.83
b1355_b_1b	6	4	10	7.33	0.73	3.67
b1355_b_2	61	24	333	10802.24	32.44	1946.35
b1908_a_1a	33	20	249	237.69	0.95	30.55
b1908_a_1b	35	23	292	804.51	2.76	93.68
b1908_b_1a	6	5	15	12.41	0.83	4.14
b1908_b_1b	6	5	14	11.80	0.84	4.21
b2670_a_1	2	2	1	1.29	1.29	1.29
b2670_b_1	2	1	1	0.73	0.73	0.73
b2670_c_1	5	4	8	8.60	1.08	4.30
b2670_d_1	2	2	1	1.27	1.27	1.27
b2670_e_1a	2	2	1	1.26	1.26	1.26
b2670_e_1b	2	1	1	0.73	0.73	0.73
b3540_a_1a	15	7	30	503.82	16.79	235.12
b3540_a_1b	10	5	19	30.50	1.61	14.45
b3540_b_1a	36	20	277	611.32	2.21	77.24
b3540_b_1b	10	5	19	31.56	1.66	14.95
b5315_a_1a	6	2	5	9.01	1.80	9.01
b5315_a_1b	3	2	2	4.37	2.19	4.37
b5315_b_1a	3	3	3	8.50	2.83	5.67
b5315_b_1b	6	2	5	8.94	1.79	8.94
b6288_a_1a	5	5	10	3720.46	372.05	1488.18
b6288_a_1b	77	42	896	5652.84	6.31	479.48
b6288_b_1a	32	19	202	1082.50	5.36	166.13
b6288_b_1b	22	16	126	825778.35	6553.80	137629.73
b7552_a_1a	13	9	43	208.98	4.86	58.32
b7552_a_1b	13	9	43	196.71	4.57	54.90
b7552_b_1	13	12	73	583.80	8.00	95.97
b7552_c_1a	16	6	25	94.81	3.79	56.89
b7552_c_1b	14	7	29	129.96	4.48	58.26

ITBS-Path was given an initial time limit of 5 seconds, which was subsequently doubled in subsequent ITBS rounds. Variables were labelled as described by simply choosing the relevant PIs with no value ordering heuristics (value 0 was tried first, then value 1). This simple strategy was sufficient to solve all 51 benchmarks over the 10 ISCAS circuits, i.e. to differentiate all the 4558 pairs that were needed to compute the 410 diagnostic partitions.

During our tests, we noticed that after running some benchmarks, the execution time increased substantially, eventually forcing us to restart SICStus. It seems that some dummy constraints remain after a differentiation attempt, and we suspect that garbage collecting might not be functioning correctly. Therefore, execution times may be somewhat inflated, especially for benchmarks where there are too many candidate diagnoses, in which case the process could not be interrupted.

Although most of the times the differentiation effort was completed in little time, some circuits had harder problems to solve. In *c1355* among the small circuits, and in *c6288* among the large ones, more difficulties were experienced and the average differentiation time was much higher. The particular benchmark *b6288_b_1b* proved to be especially hard to solve. Often, if some differentiation in a benchmark takes much longer time to solve due to its high complexity, the average differentiation time consequently increases significantly.

Of course, some variation in the results was expected, as the worst case is exponential in the number of PIs. Regardless of that, it may well be the case that obtaining a differential test pattern is easier than obtaining a simple test pattern. The reason is that a differential test pattern is more constrained than a simple test pattern and this may aid the phase of enumeration of the variables. Hence, although some differential test pattern for a pair of diagnoses may be obtained in a small amount of time, a simple test pattern for each of the diagnoses can be harder to obtain.

From the results obtained, there seems to be no significant overhead for double fault diagnoses in relation to single faults. In fact, for the circuits with double fault benchmarks, one can easily find situations where single faults differentiation takes a higher average time. Hence, the adopted approach also proves to be suitable for multiple fault differentiation.

Although the quality of results is also dependent on the type of circuit (where specialised heuristics could make a difference), we note that the average differentiation time does not explode combinatorially with the number of gates in the circuit. We thus conclude that, tipically, the complexity of the problem instances is not exponentially correlated with its size, for a moderately large number of gates (tens of thousands). Of course, some instances may be in the phase transition [Cheeseman *et al.* 1991] zone (far from trivially satisfied and far from trivially impossible), consequently being hard problems, which may happen with any (large enough) number of gates.

4.9.4 Comparison of Results and Approaches

The results obtained in the previous sections are hard to compare since existing approaches are very specific and specialised. In general, differentiation or diagnosis information is computed during simple TG in an ATG system, or derived from the resulting test set.

In [Gruning *et al.* 1991] the authors use previously generated test patterns (for SSFs in the ISCAS circuits) with the corresponding set of detected faults, to easily conclude that some pairs of faults belong to different equivalence classes (e.g. if we know that a generated test I detects a set of faults including f_1 but without f_2, then I is a differential test for f_1 with respect to f_2). After inspecting the whole test set, only a fraction of fault pairs remain undistinguished (often, mostly effectively equivalent, i.e. undistinguishable). For such remaining pairs, a diagnostic tool,

DIATEST, tries to differentiate them by also generating different output values with the aid of a 9-valued logic for the two separate faulty circuits (i.e. each with a 3-valued $\{0,1,x\}$ logic). Additionally, if a differential test pattern is found then it is added to the test set and more fault pairs may now be distinguished. Hence, fault dropping is then performed to discard those pairs already distinguished.

This is in contrast to our approach that clearly separates TG from DTG, for handling more generalised problems. Furthermore, our model is not restricted to SSFs, but rather accepts, with equivalent complexity, MSFs.

In spite of these differences we show in Table 4.24 the DIATEST results presented in [Gruning *et al.* 1991], where columns have the following meaning:

#F	number of faults to be detected
#TP	number of test patterns (in test set of ATG)
#udfp	number of undistinguished fault pairs (by the test set)
#dp	number of additional diagnostic test patterns
#dfp	number of distinguished fault pairs
#equ	number of equivalent pairs of faults identified by DTG process
T	CPU seconds for a Siemens WS30-605 workstation for DIATEST in C under UNIX

Table 4.24. DIATEST results

	#F	#TP	#udfp	#dp	#dfp	#equ	T
c432	524	59	37	13	24	13	81
c499	758	62	49	17	37	12	8
c880	942	70	55	0	0	55	1
c1355	1574	99	795	7	55	740	59
c1908	1879	118	317	16	22	295	40
c2670	2595	164	590	30	122	468	200
c3540	3428	189	577	32	46	531	397
c5315	5350	153	519	34	72	447	118
c6288	7744	40	1165	32	152	1013	259
c7552	7548	231	1261	43	143	1118	721

DIATEST thus runs on a more efficient environment. Moreover, pre-processing and specialised heuristics were used to faster find differentiating patterns. Nevertheless, the results are still comparable with ours, since we verify that most of the pairs to differentiate are really indistinguishable, in which situations we generally obtain very fast results. Equivalent faults turned out to be easier to prove since constraint propagation generally was sufficient to conclude that no *d*-signal was possible at the output. Results of Table 4.23 (previous section) also show evidence for this, since when there is a single class ($NC{=}1$, all faults equivalent) the average time is always the lowest for the benchmarks of that circuit. There is also a trend to obtain better results when NC is lower.

For a more rigorous comparison of results, we should know the exact number of DTG tries that DIATEST had to perform for each circuit and calculate the average DTG time (to compare with Ta of Table 4.23). However, since neither this number nor the number of equivalence classes are given, it has to be estimated by the provided data.

For each circuit, at least *#dp* differentiations were executed, since each such additional pattern had to be generated. This number (*#dp*) corresponds to the successful differentiations. As to the number of failed ones (proving equivalence of faults), only the number *#equ* of equivalent *pairs* of faults identified is given, which does not necessarily correspond to the number of DTG tries. In fact, if *n* faults are equivalent, then there are $n.(n\text{-}1)/2$ pairs of equivalent faults, but only *n*-1 DTG tries are required. Thus the shown number (*#equ*) may be much larger than the required number of DTG tries to prove it. Only in the very unlikely worst case, where each of the *#equ* equivalent pairs forms a distinct class, exactly *#equ* DTG tries are required to prove the equivalences. Hence, to estimate the total number of differentiation tries (ND) of DIATEST we calculate its range with a maximum given by #equ and a minimum held for n equivalent faults (n-1 DTG tries) given by the positive solution to #equ = $n.(n\text{-}1)/2$, equivalent to n^2 - n - 2.#equ = 0, yielding n = $(1+\sqrt{1+8.\#equ})/2$, for a minimum ND = n-1. Minimum and maximum average differentiation time (Ta) are then given by dividing total time (TT) by, respectively, the maximum and minimum ND.

Table 4.25 present the calculated ranges for DIATEST, together with overall results for each circuit with our approach, aggregating all single and double faults' benchmarks by summing the numbers of differentiations (ND) and total time (TT). Also for comparison we calculated the respective number of undistinguished pairs of faults (#$udfp$) given by the summation of $N*(N\text{-}1)/2$, for each benchmark of N diagnoses.

Table 4.25. Comparison of DTG results

	CLP - 8V				DIATEST			
	#udfp	ND	TT	Ta	#udfp	ND	TT	Ta
c432	32	24	25.3	1.1	37	18...26	81	3.1...4.5
c499	1188	1097	1836.0	1.7	49	22...29	8	0.3...0.4
c880	5113	448	190.3	0.4	55	10...55	1	0.0...0.1
c1355	2851	599	26619.4	44.4	795	45...747	59	0.1...1.3
c1908	1153	570	1066.4	1.9	317	40...311	40	0.4...1.0
c2670	15	13	13.9	1.1	590	61...498	200	0.7...3.3
c3540	825	345	1177.2	3.4	577	65...563	397	0.3...6.1
c5315	36	15	30.8	2.1	519	64...481	118	0.3...1.9
c6288	3663	1234	836234.2	677.7	1165	77...1045	259	0.3...3.4
c7552	445	213	1214.3	7.1	1261	90...1161	721	0.6...8.0

Actually, real average times of DIATEST are probably closer to the maximum of its range, since most pairs are in fact equivalent. For instance, for *c880* all 55 undistinguished pairs turned out to be equivalent; thus there is a single equivalence class of n faults, where $n.(n\text{-}1)/2$ = 55, which means n=11 and only 10 DTG tries are necessary. Therefore, Ta = TT/ND = 1/10 =0.1 seconds (precisely the maximum), which we can directly compare with our benchmarks where NC=1 (single class, all diagnoses equivalent). In *c880*, only *b880_b_1b* consists of just equivalent faults, with an average (and total) differentiation time of 0.29 seconds, which unfortunately is not very representative since a single differentiation try is needed for the 2 faults of the benchmark.

Nevertheless, comparing for all circuits, our results show to be competitive with those of DIATEST, with the exception of *c1355* and *c6288* where particularly hard instances pushed the average time way beyond DIATEST's ranges. In spite of such disparities, overall results are thus quite encouraging in general especially when we consider that the experimental version of our approach still has a lot of room for improvements, namely by incorporating DIATEST features such as extensive pre-processing, specialised heuristics and a much more efficient platform, and also by reducing the 8-valued logic to 4-valued for DTG (as discussed in section 4.5). Moreover, our approach also differentiates MSFs, and not just SSFs as DIATEST.

To conclude this section, we refer another approach using DIATEST that was presented in [Hartanto *et al.* 1996]. The authors also use the test set generated by the ATG system, including redundancy information, to differentiate SSFs or prove equivalence. In addition, with such information, DIATEST code was adapted to perform some local circuit transformations by a structural analysis. Equivalence of a number of faults is then easily proved. However, such fault pairs are restricted by a notion of distance that must be short to keep the problem tractable. Hence, the approach is particularly efficient for 'close' fault pairs, precisely where constraint propagation easily reaches a result.

4.10 Conclusions

In this chapter we presented an 8-valued logic and showed how it could be used to differentiate between alternative diagnoses in the cases where the circuit is not fully observable. The modelling relies on the extension of the Boolean 0/1 domain to other values that encode the dependency of these values on the alternative diagnoses. The methodology is illustrated in the problem of generation of differential test patterns in combinational circuits. Contrary to other techniques that aim at ranking candidate hypotheses based on some probabilistic indicator, or that aim at finding candidates that satisfy certain criteria based on the available information, e.g. minimal number of faulty components, our technique proves whether hypotheses are differentiable, and indicates what information is necessary to differentiate the candidates.

This problem presents a rather large search space, and to solve it efficiently we developed a specialised constraint solver for this 8-valued logic that showed to be quite effective to handle these differentiation problems.

The disjunctive nature of some constraints in the problem pushed to the limit the usual techniques to handle these constraint-solving problems, namely the cardinality operator. We therefore used the new technique, iterative time-bounded search (section 3.6), that aims at avoiding the dramatic consequences of bad initial choices in the depth-first search used in the labelling phase of constraint solving.

We applied this system in a large set of benchmarks that we created based on the standard ISCAS digital circuits, and the results obtained are quite satisfactory. In particular, they improve significantly the results obtained with a previous system that, despite being inspired by the same 8-valued logic, was implemented with a "traditional" 0/1 Boolean solver with choice points to select a sensitised path (only about 60% of the benchmarks could then be solved and quality of results was much dependent on luck and on the type of circuit — in fact, for circuits *c499*, *c3540* and *c7552*, no benchmark was solved at all).

Comparisons of our results with those obtained in the ECAD area [Hartanto *et al.* 1997, Gruning *et al.* 1991] are hard to make since we considered DTG as an independent problem (i.e. we do not take into account previous results from an ATG system). Our results are therefore more exhaustive and push to the limit the DTG problem. Still, with due distances we verified that results are competitive. Our modelling technique, by using only one circuit rather than duplications of the circuit, is competitive with current approaches in this area, even if requiring some tuning of the existing constraint solver. Of course, better results would be obtained if using a 4-valued logic (as discussed in section 4.5) with more basic gate constraints (not just 'and', 'not' and 'xor') and specialised heuristics over a more efficient platform than Prolog. But as previously stated in this thesis, our goal is not to solve the industry problem but (in addition to theoretical results) to propose new generalised efficient tools and models that the industry may use to, together with their tools and know-how, improve their results.

An interesting result of the work presented in this chapter lies in the possibility of extending our modelling to other TG problems, such as optimisation problems that will be described in the next chapters (e.g. generation of test patterns that detect a maximum number of faults).

Chapter 5

Problems with Multiple Diagnoses

In this chapter we address the issue of modelling general digital circuit problems with an arbitrary number of diagnoses. As in previous chapters, a diagnosis F is a set of faults, where each fault is a gate g stuck-at-0 ($g/0$) or stuck-at-1 ($g/1$).

Among these problems, in the ATG context, lies the generation of minimal sets of test patterns, and the related problem of finding maximal test patterns, i.e. those that maximise the number of faults they unveil. The satisfaction problem of differential diagnosis aims at generating patterns for a circuit that would entail different outputs for different sets of faulty gates, and has a possible application on the related optimisation problem of reaching the maximal fault resolution, by partitioning sets of diagnoses into equivalence classes of faults. These different problems have usually been handled either by specific tools or by modelling them in some appropriate form to be subsequently dealt with by a general problem solver (e.g. a propositional Boolean SAT-based tool [Silva *et al.* 1999]). This approach, however, requires the consideration of substantial duplication of the circuits to model diagnostic and optimisation problems [Hartanto *et al.* 1997, Gruning *et al.* 1991, Pomeranz and Reddy 1998, Manquinho and Silva 2000], which is inadequate, in practice, for a large set of diagnoses.

As an alternative to Boolean satisfiability, in the context of Constraint Logic Programming, the use of extra values in the digital signals (other than the usual 0/1) was firstly proposed [Simonis 1989] to solve the basic test generation problem and has been explored at some length in the previous chapters. In particular, we proposed an 8-valued logic to solve the differentiation problem, by introducing values that not only denote dependency on one faulty gate, but also discriminate the dependencies between the two sets of faulty gates. In this chapter we show how to associate dependency on a set of diagnoses to each expected digital signal by means of sets, which further generalises the 8-valued logic. Such extension enables the handling of satisfaction and optimisation problems with multiple diagnoses with multiple faults. The generalised technique trades off the extra circuitry required by the SAT related models with the extra values to denote the physical 0/1 Boolean values. Such richer domains require the specification of specialised logics, as well as constraint solvers to handle them efficiently.

This chapter focuses on modelling a number of problems involving multiple diagnoses, adopting a set-based representation. In the first two sections we describe such problems, considering first the satisfaction problems and then the corresponding optimisation problems. In section 5.3 we present a general logic over Booleans and sets (to express dependency on diagnoses), in order to model the different problems. Such logic is then used in section 5.4 to model and solve them. Modelling with such logic is afterwards reduced to reasoning on a set algebra in section 5.5, by means of a transformation of signal values (composite logic values over Booleans and sets) into sets alone. Conclusions are summarised in section 5.6.

5.1 Satisfaction Problems

In this section we define six satisfaction problems for some circuit under consideration. In our scope, there are three parameters involved: input and output vectors and possible diagnoses.

Problems generally consist of finding one (or a set of) such parameter given the others. The first problem, *fault simulation*, is the only one that aims at finding an output vector, which is the simplest situation since gates (including S-buffers — see section 2.5) behave as logic functions, which means that inputs determine the output, whereas the inverse is not true (an output value may be explained by more than one input combination).

5.1.1 Fault Simulation

The fault simulation problem consists of obtaining the circuit output vector, when subject to a set of faults.

- **Fault Simulation.** Find output vector o for input pattern i, of circuit with faults F.

$$fault\_simulation(i, F, o) \Leftrightarrow o = Z_F(i)$$

5.1.2 Test Generation

Let n_i be the number of PIs, I the set of all possible input vectors ($\#I=2^{n_i}$), and, using the notation of Chapter 3, $Z(i)$ and $Z_F(i)$ the output vectors under input i (where $i \in I$) for, respectively, the normal circuit and the faulty circuit exhibiting set of faults F. With such notation, a number of diagnosis related problems can be formulated, with the TG problem as the basic one:

- **Test Generation (basic).** Find an input test pattern i for a set of faults F, i.e. an input vector for which some output bit of the circuit is different when the circuit has faults F or has no faults (being thus dependent on whether the diagnosis F is correct or not).

$$test(F, i) \Leftrightarrow Z_F(i) \neq Z(i)$$

Thus *test(F,t)* means that t ($t \in I$) is an input test pattern for diagnosis F. This extends definition 3.1 to Multiple Stuck Faults (MSFs) since a diagnosis is a set of SSFs.

With the 4-valued logic, a test pattern is an input vector of the circuit that yields a PO with one extra (non Boolean) value (*d-0* or *d-1*).

Often, one is not interested in the basic problem but rather in some related and more complex problems. For problems with more than one set of faults, let D denote a set of diagnostic sets of interest (D can be a set of most common diagnoses but, in the limit, D may represent all possible sets of faults in the circuit).

Below we describe four more types of diagnostic-related satisfaction problems concerning a given set of diagnoses D: a TG generalisation and a converse problem, the diagnosis problem itself, and the differentiation problem concerning two diagnoses (i.e. $\#D=2$). Afterwards, in section 5.2, we describe two related optimisation problems where D is of arbitrary cardinality.

5.1.3 Fault Covering

A related, more general, TG problem is the problem of finding a set of test patterns for a set of diagnoses:

- **Fault Covering**. Generate a set S of input test patterns that cover all diagnoses in D.

$$cover(D, S) \Leftrightarrow \forall F \in D, \exists i \in S : test(F, i)$$

This means that each diagnosis F ($F \in D$) is detected by at least one test pattern i ($i \in S$). Set S is thus said to cover diagnoses D.

Fault covering can be naïvely solved by separately considering each diagnosis of D to find a test for it, and afterwards collect all generated tests. The number of tests is then the number of diagnoses. However, an obvious improvement consists in checking for each generated test, whether it detects other remaining diagnoses so that these can be discarded (fault dropping).

5.1.4　Covered Diagnoses

A converse problem consists of finding the set of diagnoses detected by a given input vector:

- **Covered Diagnoses**. Find diagnoses D_i, a subset of D, detected by input test pattern i.

$$covered(i, D, D_i) \Leftrightarrow D_i = \{F : F \in D \wedge test(F, i)\}$$

Test i thus covers diagnoses D_i, i.e. *cover*$(D_i, \{i\})$.

5.1.5　Diagnosis

Diagnosis as a satisfaction problem involving many possible faults, consists of obtaining sets of faults (referred to as diagnoses or diagnostic sets) that explain the circuit output for some given input vector.

- **Diagnosis**. Find the set of all diagnoses S ($S \subseteq D$) that explain circuit output vector o for input pattern i, i.e. the diagnostic sets of D that entail circuit output o under input vector i.

$$diagnosis(i, o, D, S) \Leftrightarrow \forall_{F \in S}, Z_F(i) = o \wedge \forall_{F \in D \setminus S}, Z_F(i) \neq o$$

Thus *diagnosis(i,o,D,S)* means that S are the diagnoses of D that explain circuit output o under input i.

5.1.6　Fault Location

When there are competing diagnoses that explain the faulty behaviour of a circuit, one is faced with the fault location problem that ultimately aims at finding the correct diagnosis. Such goal, however, is limited by the maximal fault resolution of the system (as explained in Chapter 4). While one can find differentiating patterns for the remaining diagnoses, one is able to improve the fault resolution. We can define this *differential diagnosis* (see previous chapter) or, simply, differentiation problem between two diagnoses as:

- **Differentiation**. Find an input test pattern i that differentiates two diagnostic sets F and G ($D = \{F,G\}$), i.e. an input vector i for which some circuit PO assumes different values when the circuit has faults F or G

$$diff(\{F, G\}, i) \Leftrightarrow Z_F(i) \neq Z_G(i)$$

This latter problem can be formulated as having some output bit dependent on a subset of D of cardinality 1. With the 8-valued logic, one aims at finding an input vector to the circuit that entails a PO containing one of the extra values that discriminate between the two diagnostic sets (*d*-signal d_1-0, d_1-1, d_2-0 or d_2-1).

5.2 Optimisation Problems

The four optimisation problems that we next describe deal with diagnostic sets D with varying cardinality.

5.2.1 Minimal Set of Test Patterns

A fault-covering related problem is the problem of minimal sets of test patterns, i.e. to generate sets of test patterns with minimum cardinality that cover all the possible faults in a digital circuit. Or, in general, for sets of faults:

- **Minimal Set of Test Patterns**. Generate a minimal set S of input test patterns that cover all diagnoses in D (in the problem definition below, $\mathcal{P}(I)$ is the power-set of I).

$$min\_cover(D,S) \Leftrightarrow cover(D,S) \wedge \forall S' \in \mathcal{P}(I), (cover(D,S') \Rightarrow \# S' \geq \# S)$$

This minimisation problem is a typical set-covering problem: the test patterns (i) are the resources, and the diagnoses (F) are the services to cover with the minimum of resources. Each diagnosis can be tested by a number of test patterns, and each test pattern can test a number of diagnoses. The relation between these services and resources is $test(F,i)$. This relation is not fully known *a priori*, though. The ultimate goal of a typical ATG system is precisely to find a set S of tests such that $min\_cover(D^{SSF},S)$ holds, where D^{SSF} is the set of all possible SSFs in the circuit.

5.2.2 Maximal Test Patterns

A problem related to covered diagnoses is the problem of maximal test patterns, i.e. to find a test pattern that maximises the number of faults it unveils. Generalising for sets of faults, we define this optimisation problem as follows:

- **Maximal Test Patterns**. Find an input i that is a test pattern for a maximum number of diagnoses in D.

$$max(D,i) \Leftrightarrow covered(i,D,D_i) \wedge \forall j \in I, (covered(j,D,D_j) \Rightarrow \# D_j \leq \# D_i)$$

The goal is to maximise the number of output dependencies on the entire set of circuit output bits, which means the number of diagnoses covered by the input test pattern i.

5.2.3 Minimal Diagnosis

Diagnosis is usually employed as an optimisation problem in that only minimal diagnoses are wanted. We define the problem using the set cardinality function to compare diagnoses:

- **Minimal Diagnosis.** Find the set of all minimal (smallest) diagnoses S ($S \subseteq D$) that explain circuit output vector o for input pattern i, i.e. the minimal diagnostic sets of D that entail circuit output o under input vector i.

$$diag_{min}(i,o,D,S) \Leftrightarrow diagnosis(i,o,D,S') \wedge S = \{F \in S': \neg \exists_{G \in S'}, \# G < \# F\}$$

A function other than set cardinality can of course be used to express a different preference of diagnoses.

5.2.4 Maximal Fault Resolution

For fault-location, we define the problem of achieving the maximal fault resolution for a set of diagnoses, as partitioning it into equivalence classes of diagnoses:

- **Classes.** Find the partition set P of sets (classes) of equivalent (indistinguishable) diagnoses in D.

$$classes(D,P) \Leftrightarrow \bigcup_{F \in P} F = D \wedge \forall_{F,G \in D}, (\neg \exists_{i \in I}, diff(\{F,G\},i) \Leftrightarrow (\exists_{C \in P}, \{F,G\} \subseteq C))$$

Two diagnoses of D either are equivalent and belong to the same class in P, or are differentiable and belong to different classes.

Some ATG systems also heuristically try to approximate this goal for all SSFs. An optimal test set in that sense would thus be the set of each test i used to differentiate two classes of P in $classes(D^{SSF},P)$.

5.3 Logic over Booleans and Sets

We now present an alternative to model the previous problems, representing digital signals with sets and Booleans. Signal values for the normal circuit and for the faulty circuits, corresponding to each possible diagnosis, can be compactly represented, in each circuit line, with a data structure consisting of a set of diagnoses and a Boolean value. This is equivalent to the deductive fault simulation technique [Armstrong 1972, Godoy and Vogelsberg 1971], which uses such a data structure to simulate the signal value in the good circuit and deduce the values in the possible faulty circuits.

This deductive modelling generalises the 8-valued logic of the previous chapter by considering more than two alternative diagnostic models.

5.3.1 Signal Representation

A digital circuit is composed of gates performing the usual Boolean operations (e.g. *and*, *not*, *xor* and simple *buffers*) and, physically, any signal in the circuit can only have a value 0 or 1. At each point, it either has the expected value (i.e. the same value of a normal circuit), or the opposite (its negation) due to the circuit fault(s) affecting it. Since the faulty behaviour can be explained by several of the possible faults, we represent a signal not only by its normal value but also by the set of diagnoses it depends on. More specifically, a signal is denoted by a pair L-N, where N is a Boolean value (representing the Boolean value of the circuit if it had no faults) and L is a set of diagnostic sets, that might change the signal into the opposite value. For instance, $X=\{\{f/0,g/0\}, \{h/1\}\}$-$0$ means that signal X is normally 0 but if both gates f and g are stuck-at-0, or if gate h is stuck-at-1, then its actual value is 1. Thus $\emptyset$-N represents a signal with constant value N, independent of any fault under consideration.

As in previous chapters, in our circuit model, we assume that there are normal gates and S-buffers (for those gates included in at least one of the possible diagnoses). We next show how the different types of gates are modelled to process signals in the form L-N.

5.3.2 Normal Gates

Normal gates (those with no faults included in D) fully respect the Boolean operation they represent. We discuss the behaviour of *not-* and *and*-gates as illustrative of these gates. The other types of gates can be modelled as combinations of these.

Given the above explanation of the encoding of digital signals, for a normal *not*-gate whose input is signal L-N, the output is simply L-$\overline{N}$, since the set of faults on which it depends is the same as the input signal. An abnormal input (due to a diagnosis in L) is a necessary and sufficient condition for an inverted output.

For the *and*-gate, three distinct situations may arise as illustrated below:

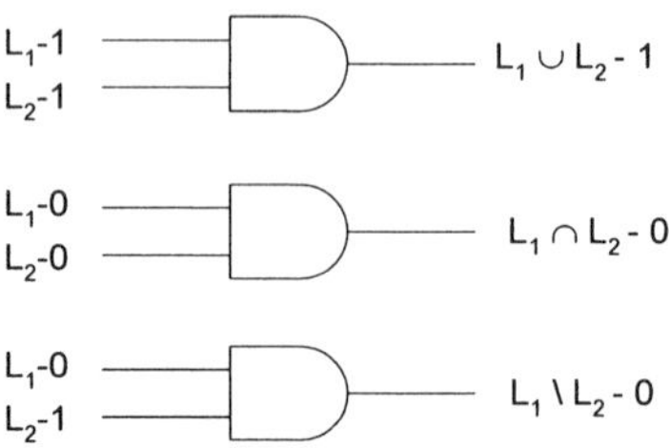

Figure 5.1. *And*-gate

In the absence of faults the output is the conjunction of the normal inputs. The set of faults on which the output is different from this normal value, is however less straightforward to determine:

1. In the first case of Figure 5.1, with two 1s as normal inputs, it is enough that a fault (more exactly, a diagnosis) in either set L_1 or L_2 occurs (toggling one input to 0) for the output to change (from 1 to 0), thus justifying the disjunction of the sets in the output signal.
2. In the second case (two 0s), it is necessary that faults occur in both L_1 and L_2 to invert the output signal, thus imposing an intersection of the input sets.
3. In the last case, to obtain an output different from the normal 0 value, it is necessary to invert the normal 0 input (i.e. to have faults in set L_1) but not the normal 1 input (i.e. no faults in set L_2), which justifies the set difference in the output.

5.3.3 S-Buffers

S-buffers are placed at the output of gates suspected to be faulty (in the universe of the possible diagnoses D). So, an S-buffer for gate g has associated a set of the possible diagnoses (diagnostic sets) where g appears as stuck. We call this set L_S. Since g can appear either as stuck-at-0 or stuck-at-1, we split this set in two (L_{S0} and L_{S1}), one for each type of diagnoses, as below:

$$L_{S0} = \{diag \in D: g/0 \in diag\}$$
$$L_{S1} = \{diag \in D: g/1 \in diag\}$$
$$L_S = L_{S0} \cup L_{S1} \quad (\text{of course, } L_{S0} \cap L_{S1} = \varnothing)$$

Only S-buffers can be stuck, and gate g is considered as normal. The modelling of S-buffers is shown in Table 5.1 for all the possible inputs to the S-buffer:

Table 5.1. S-buffer output

In	Out
$\varnothing$-0	L_{S1}-0
$\varnothing$-1	L_{S0}-1
L_i-0	$L_{S1} \cup (L_i \setminus L_{S0})$ - 0
L_i-1	$L_{S0} \cup (L_i \setminus L_{S1})$ - 1

When the input is constant 0 (thus independent of any diagnosis), the S-buffer output would normally be also 0. However, it depends on L_{S1}, because in the presence of a diagnosis where the gate is stuck-at-1, the output is 1 regardless of the input. More generally, if the normal input is 0 but dependent on L_i, the output is still dependent on L_{S1} for the same reason, but also on the input dependencies L_i that do not include fault $g/0$. This is because if the input is wrong (i.e. 1, due to Li), the output is also 1 except when it is stuck-at-0. The same reasoning can be applied to the case where the normal input signal is 1, and the whole Table 5.1 can be generalised as shown in Figure 5.2.

$$L_i\text{-N} \quad \overset{L_s}{\longrightarrow} \quad L_{S\tilde{N}} \cup (L_i \setminus L_{SN})\text{ - N}$$

Figure 5.2. S-buffer

5.4 Modelling and Solving

In this section we discuss how to model and solve the different satisfaction and optimisation problems presented in sections 5.1 and 5.2, using the logic of the previous section.

Fault simulation for an arbitrary number of diagnoses can be performed efficiently and simultaneously, since considering D as the diagnoses to simulate, inspection of circuit output with this logic trivially deduces the values for each diagnostic set F in D. In fact, for each PO b with logic value $L_b\text{-}N_b$, the simulated value is N_b if F is not a member of L_b (it is N_b for any diagnosis in $D \setminus L_b$), and its opposite otherwise (as for any diagnosis in L_b).

To model the basic *TG problem* for faults F, we would simply have to assure, for $D=\{F\}$, that some signal $\{F\}\text{-}N$ is present in a circuit output bit.

The union $\cup_b L_b$ where b ranges over all circuit output bits b with signals $L_b\text{-}N_b$, gives the *covered diagnoses* of a test, since PO b is dependent on L_b.

5.4.1 Diagnosis

In general, the universe of possible diagnoses D has to be defined. In minimal diagnosis, if only SSFs are considered, then any gate (or, more generally, signal line) may be considered as potentially faulty. D is thus the set of all such possible SSFs and each circuit gate is modelled with an additional S-buffer.

It is obviously possible to consider MSFs, since our model is general enough for that. However, considering all possible MSFs may become impracticable for very large circuits, since its number increases exponentially with the number of faults and the size of the circuit. A subset of MSFs is more likely to form the universe D in such cases. Moreover, it is natural to address first SSFs alone, and only if there is no solution, consider MSFs with an increasing number of faults.

To solve the diagnosis problem, the real circuit output has to be compared with the modelled one for all possible diagnoses D (with the above encoding by sets and Boolean values). For each PO, we compare the two values: the real z_{PO} with the modelled $L_{PO}\text{-}N_{PO}$. If z_{PO} is the same as N_{PO}, then diagnoses L_{PO} are discarded since they do not explain z_{PO}. Otherwise, diagnoses L_{PO} are an explanation for the difference. To obtain the possible diagnoses that explain the whole circuit output, all such diagnoses L_{PO} that explain a mismatch (a fault effect) in some PO must be intersected (diagnoses that would not entail all output values are discarded).

All possible diagnoses are thus obtained in just one step as opposed to traditional logic programming methods that rely on backtracking to obtain each possible solution (a single diagnosis) as in pure Prolog or in the HCLP approach described in the previous chapter, or more advanced techniques such as tabling [XSB 2000] or stable models [Gelfond and Lifshitz 1988].

In [Alferes *et al.* 2001] we present diagnosis results for the ISCAS *c6288* circuit (section 2.4) using four different approaches, namely backtracking, tabling, stable models and our deductive model using the set-Boolean encoding. The backtracking model defines each gate as a disjunction of a normal gate or a faulty one (with inverted output). The tabling approach is based on tabled abduction of XSB-Prolog, where faults are represented as abducible hypotheses (tabling avoids re-computation of goals on backtracking by memorising intermediate results). In stable models (or answer-set) programming, clauses are viewed as constraints on the diagnoses, each diagnosis being represented by a model of the program defined in terms of those constraints.

Diagnosis tests consisted in finding the whole set of SSFs that explained each of 32 incorrect output vectors obtained by flipping each PO normal value entailed by a fixed input vector. The total time (in seconds) for finding a total of 1420 faults (average of 44.4 per flipped PO) is shown in Table 5.2 with the different approaches (over different platforms).

Table 5.2. Total times for 32 incorrect outputs

	Time	Platform
Backtrack	133.4	SICStus 3.8.5 over Windows on a Pentium III (750 MHz, 128Mb)
Tabling	188.0	XSB-Prolog 2.2 over Linux on a Pentium III (733 MHz, 256Mb)
SModels	2640.2	SModels 2.26 over Windows on a Pentium III (750 MHz, 128Mb)
Deductive	0.8	SICStus 3.8.5 over Windows on a Pentium III (750 MHz, 128Mb)

Our approach is thus orders of magnitude faster than other logic programming techniques. Specialised systems such as DRUM-II [Fröhlich and Nejdl 1997] present no better results, taking 160 seconds (in 1997) to diagnose all SSFs for a specific output vector. In [Fröhlich 1998], the author even reports that several special-purpose diagnostic tools could not reliably detect faults in *c6288* circuit.

Also notice that our deductive approach is the only one that can, in practice, handle all *c6288* double faults (!), as we also report in [Alferes *et al.* 2001], where a compact representation for double faults is presented. Our approach can handle diagnoses with an arbitrary number of faults, although the consideration of all diagnoses with 3 or more faults is impractical due to its combinatorial explosion. In general, only reasonable-size subsets of the powerset of all faults may be considered, in practice, although diagnoses of a large number of faults may be modelled.

5.4.2 Differentiation

To model the problem of differentiating faults from sets F and G, (the universe being thus $D=\{F,G\}$ with cardinality 2), either $\{F\}$-N or $\{G\}$-N must be present in some circuit output bit. In both cases, a signal L-N must be present in the output where $\#L=1$, as it must be dependent on one diagnosis but not on the other.

Modelling the differentiation of two diagnoses F and G is thus straightforward. Nevertheless, with the set-Boolean encoding, one is not restricted to two diagnoses and may express an arbitrary number of diagnostic dependencies. When there are several possible diagnoses for a malfunctioning circuit, the set of all possible diagnoses may be considered when generating a differential test, in order to possibly eliminate more than one incorrect diagnosis with that test.

The differentiation problem may be generalised to finding an input test pattern i that differentiates some diagnostic sets from a set of diagnoses D, i.e. an input vector i for which some circuit PO does not assume the same value under all diagnoses D:

$$\mathit{diff}(D,i) \Leftrightarrow \exists_{F,G \in D}, Z_F(i) \neq Z_G(i)$$

which can be formulated as having some output bit dependent on a strict nonempty subset L of D (guaranteeing that diagnoses from L are distinguished from those in $\overline{L} = D\backslash L$).

Assuming that the procedure *Differential_Pattern(D, I, Out)* succeeds when such a pattern I is found for diagnoses D, yielding an output vector (with set-Boolean encoding) *Out*, we may now apply the algorithm of Figure 5.3 to find the correct class of diagnoses that explain the faulty circuit behaviour.

```
Function Fault_Location (In: Si);
    if Differential_Pattern(Si, I, Out) then   %several possible
        Z ← Real_Output(I)
        Poss_Diags ← ∅
        Imposs_Diags ← ∅
        for each PO do
            Out_PO = L_PO-N_PO      %(only) diagnoses L_PO change N_PO.
            if Z_PO = N_PO then Imposs_Diags ← Imposs_Diags ∪ L_PO
                    else Poss_Diags ← Poss_Diags ∩ L_PO
        end for
        Poss_Diags ← Poss_Diags \ Imposs_Diags
        return Fault_Location(Poss_Diags) %diagnose among possible
    else return Si            %indistinguishable: diagnosis found
end Function
```

Figure 5.3. Improved algorithm to obtain a set of indistinguishable diagnoses

The *Fault_Location* function receives as input a set of (*a priori*) possible diagnoses and returns the only equivalence class of diagnoses that explain the faulty circuit behaviour. If the initial diagnoses cannot be differentiated, then they belong to an equivalence class. Otherwise, the differentiating test found is applied to restrict the set of possible diagnoses, by comparing the real output for that test with the predicted one for all the current possible diagnoses — any diagnosis that does not entail the real output vector is discarded, the remaining ones are recursively given to the *Fault_Location* function until the final class is obtained.

Note that the potential gain of using this algorithm instead of *Differentiate*, presented in Figure 4.6, is high, since in the best case, with all diagnoses indistinguishable, we obtain the result in just one step (when failing to find a differential test). In the algorithm of the previous chapter, for *n* possible diagnoses, in spite of using smaller set domains, one had to try *n-1* pairwise differentiations to reach the same result. A similar potential improvement also occurs when a differentiating test is found, since more than one diagnosis can be discarded with that test (up to *n-1* diagnoses, actually). In the worst case, the *Fault_Location* algorithm of Figure 5.3 requires as many steps as the *Differentiate* algorithm of Figure 4.6.

5.4.3 Optimisation Problems

The first optimisation problem (minimal set of test patterns) is the typical set-covering problem driven by relation *test(F,i)*, between services (*F*) and resources (*i*), which is not, *a priori*, fully known. It may be considered as a meta-problem where we want to minimise a set of input vectors, each a set of PI assignments (since PIs do not depend on any diagnoses, they take two possible values either as $\emptyset$-0 or $\emptyset$-1 in the set-Boolean encoding). For such an input vector *i*, the relation *test(F,i)* holds if a PO assumes a logic value $\{F\}$-N (with $N \in \{0,1\}$). While the modelling of this minimisation problem is simple, its efficient solving is far from trivial and remains an open problem.

The goal of the maximal test patterns problem is to maximise the number of output dependencies, i.e. the number of diagnoses covered by the input test pattern. The goal is then simply *maximise* $\#(\cup_b L_b)$ where *b* ranges over all circuit output bits *b* with signals L_b-N_b.

For maximal fault resolution, when the real circuit is not available, algorithm *Classes* of Figure 4.5 can also be improved by the algorithm of Figure 5.4, which given as input a set of potential diagnoses, returns the partition set of all their equivalence classes.

```
Function Classes (In: Si);
    if Differential_Pattern(Si, I, Out) then   %differentiable.
        Cs ← {Si}      %initially, all diagnoses in the same class.
        for each PO do          % circuit POs
            Out_PO = L_PO–N_PO      % L_PO differentiable from Si\L_PO
            Cs' ← Cs            % classes to check in loop.
            Found ← ∅          % no class found so far.
            while Cs' ≠ ∅ and Found = ∅ do
                Cs' ← Cs' \ {Ds} % obtain a class Ds.
                if Ds ∩ L_PO ≠ ∅ then Found ← {Ds} % class found.
            end while
            if Found = {Ds}     % a class Ds with some of L_PO was found.
                then Cs ← Cs \ {Ds} ∪ {Ds ∩ L_PO} ∪ {Ds \ L_PO}
                    % update Cs by splitting its element Ds in 2
        end for
        Classes ← ∅               % no definitive classes yet.
        for each Set in Cs do Classes ← Classes ∪ Classes(Set)
        return Classes
    else return {Si}            % indistinguishable.
end Function
```

Figure 5.4. Improved algorithm to partition a set into classes of indistinguishable
diagnoses

The recursive function *Classes* represents an improvement over that of Figure 4.5 since when
dealing with multiple diagnoses, a differential test may differentiate more than two diagnoses.
Potentially, all n possible diagnoses can be differentiated with just a test since, for example, each
diagnosis may affect a different PO. Hence, the set of all equivalence classes may be returned
with just one step, in the best case. The same happens if all diagnoses are equivalent, being its
class returned with just one (failed) differentiation try.

5.5 Reduction to Set Algebra

Throughout this chapter we have shown how a specific signal encoding that handles an arbitrary
number of circuits (one normal and the others faulty) may be used to model a number of
diagnostic related problems. Clearly, the practical interest of this work depends on the ability to
develop adequate constraint solvers to deal with the domains that have been used (Booleans and
sets). In previous chapters we have already shown some constraint solvers over existing
Constraint Logic Programming languages (SICStus [SICStus 1995] and ECLiPSe [ECRC 1994])
to solve different problems with specific multi-valued logics. Nevertheless, no particular logic was
general enough to model the set of problems presented in this chapter, so we presented a general
logic obtained by combining two domains (Booleans and sets). However, such combination may
prevent its efficient solving, so we show in this section how these problems can be modelled and
solved with pure set reasoning. In Chapter 6 we present a general set constraint solver able to
efficiently handle such problems.

In this section we discuss a CLP approach over sets as a unifying modelling framework for all the
above presented diagnostic related problems, whereby a signal can be simply represented by a set
thus eliminating the need of a Boolean value.

5.5.1 Motivation

Modelling circuit problems with sets and Booleans involves handling disjunctions on gates since, with the signal representation described in section 5.3.1, the two domains are not independent. In fact, the output set depends not only on the input sets but also on the Boolean inputs. The relation is not easily expressed as an equation and the output set expression itself is a function of the Boolean inputs in the form of a disjunction. More specifically, the set of diagnoses of the output can either be the union, intersection or difference of the input sets, depending on the normal input Boolean values. Hence, the relation is expressed by a disjunction, which may lead to inefficient computations if the Boolean values are not known *a priori*. Moreover, a logic gate defines a relation between inputs and outputs, and several problems require, for an efficient solving, inference capabilities in both ways, i.e. from output to inputs as well as from inputs to output.

We have seen that constraints are a natural and efficient way to express logic relations for combinatorial problems. However, with two different inter-related domains (sets and Booleans), solving becomes harder since two solvers have to interact and co-operate to handle disjunctions over the two domains, thus incurring a significant overhead.

Finding an equivalent representation involving just one domain could thus significantly ease both problem modelling and solving.

5.5.2 Transformation

With the previous representation, all digital signals are represented by a pair: a set of diagnoses on which it depends plus a Boolean value that the signal takes if there were no faults at all. Both the set and the Boolean value can be variables, enforcing constraints on two domains to be expressed for each gate, and the modelling presented above implies an extensive use of disjunctive constraints, with the corresponding exponential complexity. For instance, to express the above *and*-gates, one needs to know the Boolean values of the signals to select the appropriate set constraint.

It is thus very convenient to join the two domains into a single one. Intuitively, to incorporate the two domains, the new one should be richer than any of them. But there is also the possibility of using a simpler one if the loss of information is not important for the problem, or if it can be compensated by the introduction of extra constraints. This latter alternative is the one we follow here.

More specifically, we propose the use of a transformation *transf* [Azevedo and Barahona 2000a, 2000d], where signals *L-0* are simply represented as *L*, and *L-1* as $\overline{L}$ (the complement of *L*, w.r.t the set universe domain which is the set of diagnoses *D*):

$$transf(S) = \begin{cases} L, & S = L - 0 \\ \overline{L}, & S = L - 1 \end{cases}$$

According to [Abramovici *et al.* 1990], such a transformation had already been privately proposed by Levendel [1980] for SSFs in order to represent the list of faults that cause the physical signal value of *S* to take value 1, but no further use was made of it.

Although *transf* is not a bijective function (both *L-0* and $\overline{L}$*-1* are transformed into *L*), it is quite useful to model certain digital circuits problems. This transformation is similar to the reduction of the 8-valued logic into a 4-valued logic for differentiation (described in section 4.5), in that the normal value is lost. Set *L* now represents the list of diagnoses that make the signal assume value 1, which is precisely what that 4-valued logic (for faults ϕ_1 and ϕ_2) encodes: the 4 values correspond to the power-set of the 2 diagnoses, i.e. $\{\{\}, \{\phi_1\}, \{\phi_2\}, \{\phi_1, \phi_2\}\}$.

To exemplify the usefulness of the transformation, circuit inputs which can only have value 0 or 1, if represented by set variables, can have values $\varnothing$ or D. As to logical gates, the *and*-gate and the S-buffer are simply stated as follows (Figure 5.5):

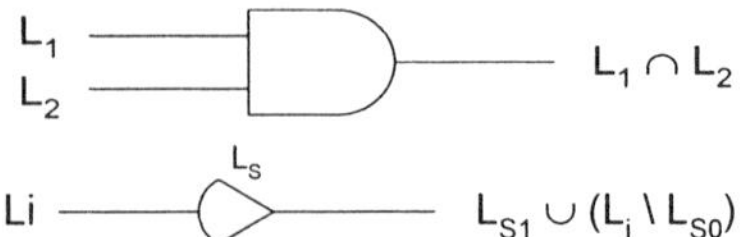

Figure 5.5. *And*-gate and S-buffer over sets

The correctness of this new simplified representation can be checked by simple analysis of each case, shown in Table 5.3 and Table 5.4.

Table 5.3. Application of *transf* function to the inputs and output of an *and*-gate

I1	*transf(I1)*	*I2*	*transf(I2)*	*I1∧I2*	*tranf(I1∧I2)*
$L_1\text{-}1$	$\overline{L_1}$	$L_2\text{-}1$	$\overline{L_2}$	$L_1 \cup L_2\text{-}1$	$\overline{L_1 \cup L_2} = \overline{L_1} \cap \overline{L_2}$
$L_1\text{-}0$	L_1	$L_2\text{-}0$	L_2	$L_1 \cap L_2\text{-}0$	$L_1 \cap L_2$
$L_1\text{-}0$	L_1	$L_2\text{-}1$	$\overline{L_2}$	$L_1 \backslash L_2\text{-}0$	$L_1 \backslash L_2 = L_1 \cap \overline{L_2}$

Table 5.4. Application of *transf* function to the input and output of an S-buffer

In	*transf(In)*	*S-buffer output*	*transf(output)*
$L_i\text{-}0$	L_i	$L_{S1} \cup (L_i \backslash L_{S0})\text{ - }0$	$L_{S1} \cup (L_i \backslash L_{S0})$
$L_i\text{-}1$	$\overline{Li}$	$L_{S0} \cup (L_i \backslash L_{S1})\text{ - }1$	$\overline{L_{S0} \cup (L_i \backslash L_{S1})} = \overline{L_{S0}} \cap \overline{L_i \cap \overline{L_{S1}}} = \overline{L_{S0}} \cap (\overline{L_i} \cup L_{S1}) =$

$$(\overline{L_{S0}} \cap \overline{L_i}) \cup (\overline{L_{S0}} \cap L_{S1}) = (\overline{L_i} \cap \overline{L_{S0}}) \cup L_{S1} =$$

$$L_{S1} \cup (\overline{L_i} \backslash L_{S0})$$

In Table 5.3, the transformed output set is always the intersection of the transformed input sets, i.e. *transf(I1)* ∧ *transf(I2)* = *transf(I1 ∧ I2)*. Similarly, in Table 5.4, if we represent the output of an S-buffer with associated dependencies L_s, subject to input i as *s_buffer(L_s,i)*, then *s_buffer(L_s,transf(Input))* = *transf(s_buffer(L_s, Input))*.

For completion, it may be also noticed that the other gate operations can be expressed with the expected set operations (Figure 5.6):

Figure 5.6. Other gates over sets

5.5.3 Modelling

To solve the differentiation problem between two diagnoses with this representation based exclusively on sets, it is still sufficient to ensure that a set L with cardinality 1 is present in a circuit PO. If $D=\{F,G\}$ is the set of diagnoses F and G to differentiate, then when a PO takes value L (with $\#L=1$) in the purely set-based representation, it is equivalent to taking value L-N (with $\#L=1$) in the mixed representation (pairs Set-Boolean).

The proof is straightforward. If set L ($\#L=1$) is present in a PO, then it represents either L-0 ($\#L=1$) or $\overline{L}$-1 ($\#\overline{L}=1$, since $\#D=2$), both sets with cardinality 1. Conversely, if a pair L-N

(#L=1) is present in a PO, it is either *L-0* (represented as set L) or *L-1* (represented as $\overline{L}$). In either case, the represented set has cardinality 1.

Therefore, the loss of information incurred by the transformation used has no effect in this problem, since it is not necessary to add any new constraints to solve it.

A similar situation occurs when trying to differentiate more than two diagnoses simultaneously (potentially more efficient). In such case, the differentiation condition for diagnoses D of arbitrary cardinality can be generalised to having a PO with set value L, such that L is neither the empty set nor the universe D (i.e. $L{\neq}\varnothing \wedge L{\neq}D$), since then it is assured that the generated test pattern differentiates (at least) diagnoses L from $D\backslash L$.

For the optimisation problem of maximal test patterns in a circuit c, the goal is to maximise the number of output dependencies, (i.e. the number of diagnoses covered by the input test pattern) which is not so straightforward to express. Since a digital signal coded as set L does not necessarily mean a dependency on diagnoses L (it can represent *L-0*, as well as $\overline{L}$*-1*), maximising the union of all the output bits is not adequate. In fact, it is necessary to know exactly whether an output signal depends on its set or on its complement. This can be done by duplicating the circuit to recover the lost information (the normal output values) as illustrated in Figure 5.7:

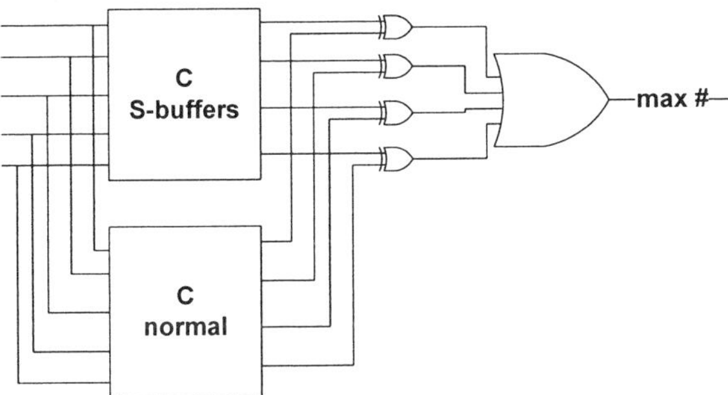

Figure 5.7. Modelling the maximisation problem with sets

Circuit c with S-buffers is kept as before, but now the circuit with no S-buffers (i.e. with all gates normal) is added, sharing the PIs and with the corresponding POs *xor*-ed. Values inside the normal circuit are necessarily independent of any faults, and can only be represented as $\varnothing$ (for $\varnothing$-0) and D (for $\varnothing$-1). The *xor*-gates in the output bits receive a set L from the faulty circuit (i.e. with S-buffers) and either $\varnothing$ or D (the universe) from the normal one. Therefore, L is kept as L if the normal value was 0, and recovered to $\overline{L}$ if the normal value were 1, just like our *transf* function. Hence, the correct dependency set of the signal is recovered and a maximisation on the union (*or*-gate) of these real diagnostic dependencies can be performed to obtain a desired solution to the problem.

The reduction of the problem size by eliminating the Boolean part of the domain is now compensated by the duplication of constraints. Still, what could naively be seen as a useless manipulation, allows an active use of constraints by a set constraint solver avoiding the choice-points that would otherwise be necessary.

Moreover, the exponential component of search, labelling, is only performed at the circuit with S-buffers (the other circuit simply checks this labelling). This is in contrast with Boolean SAT approaches, which consider one extra circuit for each diagnosis, which is unacceptable, in practice, for a large set of diagnoses [Silva *et al.* 1999].

The optimisation problem of minimal set of test patterns is the set-covering problem where test patterns are the resources, and diagnoses are the services we want to cover with the minimum of resources. Unfortunately, we do not know *a priori* whether some input vector i covers a diagnosis F, (i.e. *test(F,i)*); otherwise we could simply apply the integer linear programming (ILP) approach to set-covering (a well studied problem with easily available specific efficient tools to solve it).

This minimisation problem is thus a meta-problem: it involves sets of solutions to set problems (TG is a set problem having as solution an input vector, i.e. a set of specified PIs, and we want to minimise the set of such solutions for all diagnoses D). A set variable S could be used ranging from $\varnothing$ to $P(I)$, where set S of inputs is constrained to cover diagnoses D. The goal is then to minimise the cardinality of S.

To find a test pattern for a single diagnosis F using sets, we need to model a faulty and a normal circuit *xor*-ing the outputs and checking whether at least one set value $\{F\}$ is obtained. This is equivalent to the SAT approach for obtaining test patterns.

The ideal is, however, to consider all diagnoses D at the same time, with set constraints, and include or remove elements from S during the computation, updating the diagnoses covered until D is reached, and then start finding smaller sets for S in a branch-and-bound manner. This is still an open problem and the maximisation problem may perhaps be used to solve this minimisation one, by obtaining intermediate solutions.

5.6 Summary

In this chapter we showed how to model diagnostic related problems in digital circuits with a constraint logic programming approach. We reckon our approach has great potential in this area, since competing alternatives, based on SAT, require substantial duplication of the circuits under consideration. In contrast, our technique uses set variables to denote dependency of faults and is able to model the problems without adding extra circuitry (more precisely, without imposing the labelling of more variables, the exponential part of search).

To conclude this chapter, we note that although avoiding the duplication of circuitry required by a Boolean approach, the domains of the variables in this new modelling become more complex. In fact, the new domains are now sets of diagnoses. A constraint solver over sets is thus required to efficiently deal with the above described set problem models directly. This is the topic of the next chapter where we describe a new such set solver, *Cardinal*.

Chapter 6

A New Set Constraint Solver: Cardinal

In the previous chapter we showed how different diagnostic related problems could be modelled using sets to denote dependency and justified the convenience of taking a CLP approach over sets to efficiently solve such problems. To deal with set variables and set constraints, we realised that existing set constraint solvers were not adequate to handle these problems, as they were not actively using important information about the cardinality of the sets, a key feature in these problems.

In this chapter we present a new general set constraint solver for combinatorial problem solving, *Cardinal*, that exploits inferences over sets cardinality, and we illustrate its efficiency with experimental results. *Cardinal* is formally presented as a set of rewriting rules on a constraint store. We then show the importance of propagating constraints on sets cardinality, by comparing *Cardinal* mainly with a simpler version that propagates these constraints similarly to *Conjunto*, a widely available set constraint solver. This simpler version, using a more limited amount of constraint propagation on cardinalities will be referred in the following simply as "Conjunto". For some applications, we also present results over real *Conjunto*. The reason why we did not always compare our results with the real version was that we wanted a fairer comparison to assess the advantages of our especial inferences, and also because some 'bugs' in the real *Conjunto* version at the time of some experiments prevented us from obtaining the correct results. (Implementation details seem to be relatively unimportant, since "Conjunto" behaved as the real *Conjunto* on a set of experiments where *Conjunto* succeeded to execute.) Results show that *Cardinal* obtains a speed up of about two orders of magnitude over "Conjunto", on a set of diagnostic problems.

This chapter is organised as follows: section 6.1 introduces and discusses approaches on set constraint solving in general and cardinality inferences in particular. Section 6.2 then reviews some basic concepts of set theory that will be useful to describe the operational semantics of *Cardinal* and its constraints in section 6.3. Implementation is briefly discussed in section 6.4 and some results on differential diagnosis presented in section 6.5 and compared with "Conjunto". Then, in section 6.6, we present other applications of *Cardinal* with more results and comparisons to other approaches. Finally, in section 6.7 we present *Cardinal* extensions on set functions, exemplifying with the union function on the set-covering problem, and discuss their potentialities and possible lines of research.

6.1 Set Constraint Solving and Cardinality Inferences

A set is naturally used to collect distinct elements sharing some property. Combinatorial search problems over these data structures can thus be naturally modelled by high level languages with set abstraction facilities, and efficiently solved if constraint reasoning prunes search space when the sets are not fully known *a priori* (i.e. they are variables ranging over a set domain).

Set constraints have deserved in the last years special attention by the Constraint Programming community and have been addressed in recent literature for set-based program analysis [Heintze and Jaffar 1994] and for general set-based combinatorial search problems [Caseau *et al.* 1999, Azevedo and Barahona 2000d]. Many interesting theoretical and practical results were obtained [Charatonik and Podelski 1996, Devienne *et al.* 1997, Gervet 1997] making it a very rich and promising research topic [Aiken 1994, Pacholski and Podelski 1997].

Many complex relations between sets can be expressed with constraints such as set inclusion, disjointness and equality over set expressions that may include such operators as intersection, union or difference of sets. Also, as it is often the case, one is not interested simply on these relations but on some attribute or function of one or more sets (e.g. the cardinality of a set). For instance, the goal of many problems is to maximise or minimise the cardinality of a set. Even for satisfaction problems, some sets, although still variables, may be constrained to a fixed cardinality or a stricter cardinality domain than just the one inferred by the domain of a set variable (for instance, the cardinality of a set may have to be restricted to be an even number).

Conjunto [Gervet 1997] was the first language to represent set variables by set intervals with a lower and an upper bound considering set inclusion as a partial ordering. Consistency techniques are then applied to set constraints by interval reasoning [Benhamou 1995]. In *Conjunto* (available as an ECLiPSe [ECRC 1994] library), a set domain variable S is specified by an interval $[a,b]$ where a and b are known sets ordered by set inclusion, representing the greatest lower bound and the lowest upper bound of S, respectively.

To deal with optimisation problems, *Conjunto* includes the cardinality of a set as a graded function in the system, and generalises a graded function as $f : \mathcal{P}(\mathcal{H}_u) \rightarrow \mathbb{N}$ mapping a non quantifiable term of the power-set of the Herbrand universe to a unique integer value denoting a measure of the term, and satisfying $s_1 \subseteq s_2 \Rightarrow f(s_1) \leq f(s_2)$ for two sets s_1, s_2.

The cardinality of a set S, given as a finite domain variable C ($\#S=C$), is not a bijective function since two distinct sets may have the same cardinality. Still, due to the properties of a graded function, it can be constrained by the cardinalities of the set bounds. *Conjunto* allows graduated constraints over cardinalities but this cardinality information is largely disregarded until it is known to be equal to the cardinality of one of the set bounds, in which case an inference rule is triggered to instantiate the set.

In the problems over digital circuits discussed in the previous chapter, PIs (which can only have value 0 or 1) if represented by set variables (S_i), can have values $\emptyset$ or D and are thus given a set interval domain $[\emptyset,D]$, whose cardinality $\#S_i$ has only the two possible integer values $\{0,\#D\}$. This cardinality information, however, is mostly ignored in *Conjunto*, as explained. We thus propose that set constraint solvers must handle the sets cardinality more actively, given the important role this feature plays in diagnostic related problems.

Hence, although *Conjunto* represented a great improvement over previous CLP languages with set data structures [Gervet 1997], it lacked some inferences on the cardinality level, which are crucial for a number of CSPs.

In fact, *Conjunto* makes a very limited use of the information about the cardinality of set variables. The reason for this lies in the fact that it, is in general, too costly to derive all the inferences one might do over the cardinality information in order to tackle the problems *Conjunto* had initially been designed for (i.e. large scale set packing and partitioning problems) [Gervet 1999]. Nonetheless, and given their nature, we anticipated that some use of this information could be quite useful and speed up the solving of these problems.

Inferences using cardinalities can be very useful to deduce more rapidly the non-satisfiability of a set of constraints, thus improving efficiency of combinatorial search problem solving. As a simple example, if Z is known to be the set difference between Y and X, both contained in set $\{a,b,c,d\}$, and it is known that X has exactly 2 elements, it should be inferred that the cardinality of Z can never exceed 2 elements (i.e. from $X,Y \subseteq \{a,b,c,d\}$, $\#X=2$, $Z=Y\backslash X$ it should be inferred that $\#Z \leq 2$). A failure could thus be immediately detected upon the posting of a constraint such as $\#Z=3$.

Inference capabilities such as these are particularly important when solving set problems where cardinality plays a special role, as is the case of the circuit problems seen above. We therefore developed a new constraint solver over sets that fully uses constraint propagation on sets cardinality.

Before discussing the solver, we will overview some theoretical concepts required for its explanation.

6.2 Intervals and Lattices

Set intervals define a lattice [Birkhoff 1967, Graetzer 1971, Gierz *et al.* 1980] of sets. Figure 6.1 illustrates the powerset lattice for the example set domain $\mathcal{U} = \{a,b,c,d\}$, where a line connecting set S_1 to underneath set S_2 means $S_2 \subseteq S_1$. The set inclusion relation $\subseteq$ between two sets defines a partial order on powerset $\mathcal{P}(\mathcal{U})$, the set of all subsets of $\mathcal{U}$. Hence $\mathcal{P}(\mathcal{U})$ is a partially ordered set where binary relation $\subseteq$ has the following properties:

- $\forall S : S \subseteq S$ **Reflexivity**
- $\forall S,T : ((S \subseteq T) \wedge (T \subseteq S)) \Rightarrow (S = T)$ **Antisymmetry**
- $\forall S,T,U : ((S \subseteq T) \wedge (T \subseteq U)) \Rightarrow (S \subseteq U)$ **Transitivity**

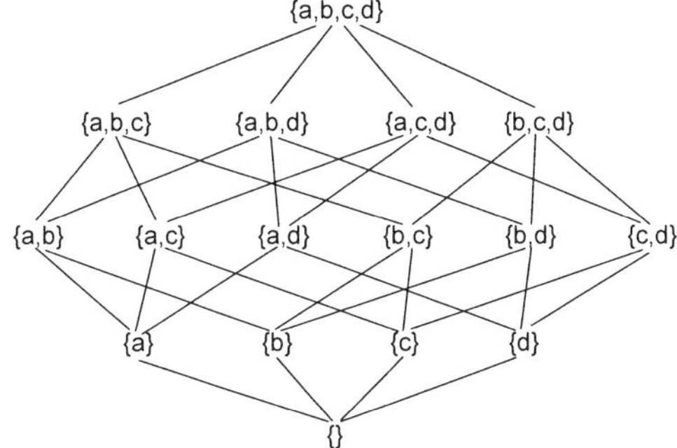

Figure 6.1. Powerset lattice for $\mathcal{U} = \{a,b,c,d\}$, with set inclusion as partial order

There are thus other implicit inclusion relations in the lattice of Figure 6.1 that were not explicitly drawn, due to the transitivity rule. The top set, $\mathcal{U}$, includes all sets of $\mathcal{P}(\mathcal{U})$; while the bottom set, $\{\}$, is included in all sets of $\mathcal{P}(\mathcal{U})$. Consequently, sets $\mathcal{U}$ and $\{\}$ constitute an upper bound and a lower bound of $\mathcal{P}(\mathcal{U})$, respectively. In addition, they are the **least upper bound** (lub) or *join*, and the **greatest lower bound** (glb) or *meet* of $\mathcal{P}(\mathcal{U})$, since there is no other upper bound contained in ('less' than) $\mathcal{U}$ nor other lower bound containing ('greater' than) the empty set $\{\}$.

Let us now consider the sub-lattice of Figure 6.2 (a). Sets $\{\}$ and $\{a,b,c,d\}$ are still a lower and an upper bound, but this time the *glb* is $\{b\}$ and the *lub* is $\{a,b,d\}$.

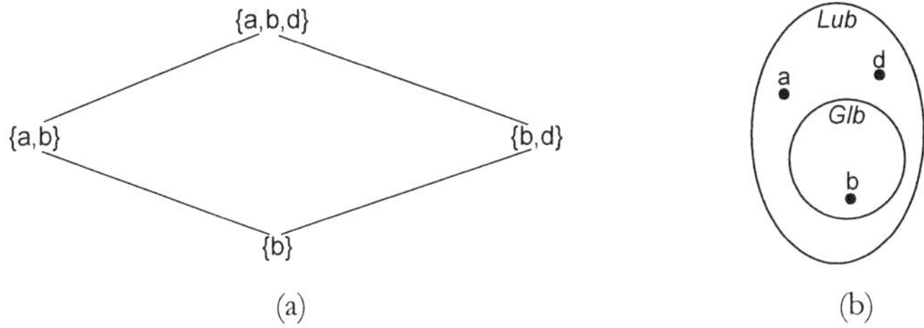

(a)   (b)

Figure 6.2. Set interval [$\{b\},\{a,b,d\}$]: a) Sub-lattice; b) Venn diagram

The two bounds (*glb* and *lub*) define a set interval (e.g. [{*b*},{*a,b,d*}]) and may form the domain of a set variable *S*, meaning that set *S* is one of those defined by its interval (lattice); all other sets outside this domain are excluded from the solution. Thus, *b* is definitely an element of *S*, while *a* and *d* are the only other possible elements (Venn diagram in Figure 6.2 (b)).

Set interval reasoning allows us to apply consistency techniques such as Bounded Arc Consistency (see section 1.2), due to the monotonic property of set inclusion.

Any set variable must then have a domain consisting of a set interval. In addition, this interval should be kept as small as possible, in order to discard all sets that are known not to belong to the solution, while not loosing any of the still possible values (sets). The smallest such domain is the one with equal *glb* and *lub*, i.e. a domain of the form [*B,B*], corresponding to a constant set *B*. For a set variable that can assume any set value from a collection of known sets, such as {{*a,b*},{*a,c*},{*d*}}, the corresponding interval is the convex closure of such collection (which in this case is the set interval [{},{*a,b,c,d*}]). In general, for *n* possible arbitrary sets $S_1...S_n$, the corresponding set variable *X* has an interval domain [*glb, lub*] where

$$glb = \bigcap_{i=1}^{n} S_i \quad and \quad lub = \bigcup_{i=1}^{n} S_i$$

In the next section we describe *Cardinal*'s propagation of constraints over set variables having such interval domains, together with inferences over sets' cardinality.

6.3 Operational Semantics

The set universe notion is necessary not only for the set complement operation, but also for the especial cardinality inferences we propose. Hence we will use $\mathcal{U}$ to denote the set universe domain (for the circuit problems proposed in Chapter 5, the universe is the set of diagnoses *D*), and *u* to denote the cardinality of $\mathcal{U}$ ($u = \# \ \mathcal{U}$).

A set variable *X* is represented by $[a_X,b_X]_{C_X:D_X}$ (or simply as $[a_X,b_X]_{C_X}$) where a_X is its greatest lower bound (i.e. the elements known to belong to *X*), b_X its lowest upper bound (i.e. the elements not excluded from *X*), and C_X its cardinality (a finite domain variable) with domain D_X. In the remainder, a_X, b_X, C_X and D_X will be used to refer to these attributes of set variable *X* if no confusion arises.

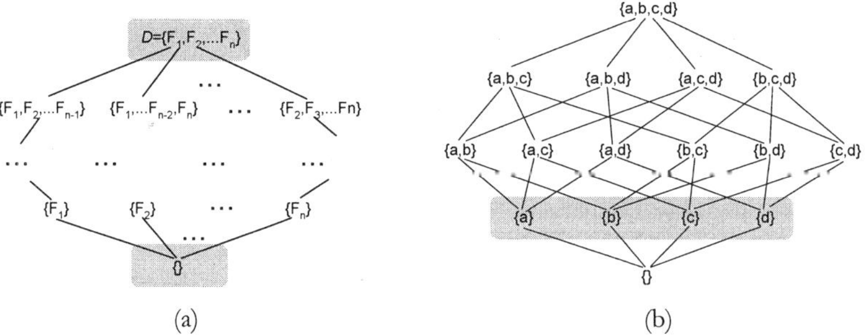

Figure 6.3. Powerset lattices with cardinalities: a) circuit PI; b) singleton

Given the transformation presented in section 5.5.2 to encode circuit signals by sets, independent values such as the circuit inputs are represented by $[\emptyset,\mathcal{U}]_{C_X:\{0,u\}}$. The only two possible values are then $\emptyset$ ($C_X=0$) and $\mathcal{U}$ ($C_X=u$), shown in gray in Figure 6.3 (a). We may view powerset lattices as a number of horizontal layers of different cardinality; the set cardinality variable constrains possible layers where the solution value is. Figure 6.3 (b) shows the lattice for set variable *X* with domain [{},{*a,b,c,d*}], corresponding to a singleton. Such domain is enough to express the disjunction

$X=\{a\} \vee X=\{b\} \vee X=\{c\} \vee X=\{d\}$ and, therefore, one does not need to explicitly post this disjunctive constraint.

Cardinal implements a number of set constraints such as inclusion, equality, inequality, membership and disjointness, together with set operations (union, intersection, difference and complement), as built-in. (Nonetheless, the equality and inclusion constraints together with the operations of sets complement and binary intersection are sufficient to model circuit problems.) We will next describe all these *Cardinal* constraints. Inferences will be formally described as rewriting rules as in the following schematic figure:

$$(\textit{trigger condition}) \qquad \frac{pre-conditions}{CS\_changes}$$

where CS refers to the Constraint Store. In addition, a number on the right identifies each inference rule.

The constraint store maintains constraints over sets and over finite domains (the cardinality of the sets), but we only describe the rewriting rules of the set constraints (we assume that a finite domain constraint solver maintains bounded arc-consistency, or interval consistency, on these constraints).

6.3.1 Set Variable

When a variable is declared as a set variable, it is simply included in the constraint store, ensuring the bounds of the variable and that its cardinality C is a finite domain variable with domain D:

$$\overline{\{tell(X \in [a,b]_{C \,:\, D})\} \mapsto \{X \in [a,b]_C, C :: D\}} \tag{1}$$

A number of inferences are subsequently maintained. During computation, set intervals can only get shorter, i.e. either because the lower bound becomes larger or the upper bound smaller, or both.

The cardinality must always remain inside the limits given by the set bounds (the triggers of these inferences are shown in parenthesis next to the rewriting rules, and may correspond to one or more variables becoming ground, changing bounds, or being bound in the Prolog sense):

$$(X\text{: changed bounds}) \qquad \frac{n = \#\,a_x, m = \#\,b_x}{\{\} \mapsto \{C_x \geq n, C_x \leq m\}} \tag{2}$$

As in *Conjunto*, a set variable becomes one of its bounds if their cardinality is the same (this rule is triggered only when C_X becomes a fixed value):

$$(C_X\text{: ground}) \qquad \frac{C_x = \#\,a_x}{\{\} \mapsto \{X = a_x\}} \qquad \frac{C_x = \#\,b_x}{\{\} \mapsto \{X = b_x\}} \tag{3}$$

When there are two domains declared for the same set variable, their intersection must be computed and the cardinalities made equal:

$$\frac{a = a_1 \cup a_2, b = b_1 \cap b_2}{\{X \in [a_1,b_1]_{C_1}, X \in [a_2,b_2]_{C_2}\} \mapsto \{X \in [a,b]_{C_1}, C_1 = C_2\}} \tag{4}$$

Eventually a failure may be detected, either because the lower bound of a set is not included in its upper bound, or the domain of the cardinal becomes empty:

$$(X\text{: changed bounds}) \qquad \frac{not(a_x \subseteq b_x)}{\{\} \mapsto fail} \qquad \frac{D_x = \varnothing}{\{\} \mapsto fail} \tag{5}$$

6.3.2 Membership Constraints

Membership constraint is trivially handled by inserting, as soon as it is ground, the given element in the set's *glb*:

$$\frac{\rule{0pt}{0pt}}{\{tell(elem \in X)\} \mapsto \{elem \in X\}} \tag{6}$$

$$(elem:\ \text{ground}) \quad \frac{a = a_x \cup \{elem\}}{\{elem \in X\} \mapsto \{X \in [a, b_x]\}} \tag{7}$$

Its negation is similarly handled by removing the element from the *lub*:

$$\frac{\rule{0pt}{0pt}}{\{tell(elem \notin X)\} \mapsto \{elem \notin X\}} \tag{8}$$

$$(elem:\ \text{ground}) \quad \frac{b = b_x \setminus \{elem\}}{\{elem \notin X\} \mapsto \{X \in [a_x, b]\}} \tag{9}$$

6.3.3 Set Complement

For the set complement constraint it is assumed the existence of a non-empty universe of cardinality u. Hence, a set cannot be the same as its complement:

$$\frac{\rule{0pt}{0pt}}{\{tell(X = \overline{Y}), X = Y\} \mapsto fail} \tag{10}$$

Cardinality u is used in a finite domain constraint, $C_Y = u - C_X$, over the sets cardinalities. In general, the finite domains constraint solver maintains bounded arc consistency on this constraint. Nevertheless, we ensure full arc consistency when the constraint is successfully posted:

$$\frac{\rule{0pt}{0pt}}{\{tell(X = \overline{Y})\} \mapsto \{C_y = u - C_x, X = \overline{Y}\}} \tag{11}$$

Afterwards, whenever there is an update of the bounds of one of the sets, the bounds of its complement must also be updated accordingly, which eventually leads to the successful instantiation of the sets or to a failure:

$$(X:\ \text{changed bounds}) \quad \frac{a = \overline{b_x}, b = \overline{a_x}}{\{X = \overline{Y}\} \mapsto \{X = \overline{Y}, Y \in [a, b]\}} \tag{12}$$

$$(Y:\ \text{changed bounds}) \quad \frac{a = \overline{b_y}, b = \overline{a_y}}{\{X = \overline{Y}\} \mapsto \{X = \overline{Y}, X \in [a, b]\}} \tag{13}$$

The constraint disappears (is trivially proved or disproved) when sets are ground and their complementary nature can be easily checked:

$$(X\ \text{or}\ Y:\ \text{ground}) \quad \frac{ground(X), ground(Y), a_x = \overline{a_y}}{\{X = \overline{Y}\} \mapsto \{\}} \qquad \frac{ground(X), ground(Y), a_x \neq \overline{a_y}}{\{X = \overline{Y}\} \mapsto fail} \tag{14}$$

6.3.4 Set Equality

When two sets are told to be equal, so does their cardinality:

$$\frac{}{\{tell(X = Y)\} \mapsto \{X = Y, C_x = C_y\}} \tag{15}$$

When one bound of the set is updated, so does the corresponding bound of any set equal to it (the situation is similar to that of having two domains for the same set, as in rule 4):

$$(X \text{ or } Y: \text{ch. bounds}) \quad \frac{a = a_x \cup a_y, b = b_x \cap b_y}{\{X = Y, X \in [a_x, b_x], Y \in [a_y, b_y]\} \mapsto \{X = Y, X \in [a, b], Y \in [a, b]\}} \tag{16}$$

Again, when sets are ground, the equality is easily confirmed (or infirmed):

$$(X \text{ or } Y: \text{ground}) \quad \frac{ground(X), ground(Y), a_x = a_y}{\{X = Y\} \mapsto \{\}} \quad \frac{ground(X), ground(Y), a_x \neq a_y}{\{X = Y\} \mapsto fail} \tag{17}$$

Of course, if only one of the sets becomes ground, the previous rule enforces the other set either to become with the same bounds (and ground, in which case this rule eliminates the equality constraint) or with an empty domain, causing a failure.

6.3.5 Set Inequality

When two sets are told to be different, we cannot relate their cardinalities since these can be equal, even with different sets:

$$\frac{}{\{tell(X \neq Y)\} \mapsto \{X \neq Y\}} \tag{18}$$

Of course, the two sets cannot be the same:

$$(X \text{ or } Y: \text{bound}) \quad \frac{}{\{X \neq Y, X = Y\} \mapsto fail} \tag{19}$$

The inequality constraint does not allow much propagation, hence we wait until a set (X) is ground. Then, if its cardinality cannot be the same as the other ($\#Y$), the constraint is satisfied:

$$(X: \text{ground}) \quad \frac{C_x \neq C_y}{\{X \neq Y\} \mapsto \{\}} \tag{20}$$

If X is equal to a bound of Y, then we just have to assure the cardinality is different:

$$(X: \text{ground}) \quad \frac{X = a_y}{\{X \neq Y\} \mapsto \{C_y > C_x\}} \quad \frac{X = b_y}{\{X \neq Y\} \mapsto \{C_y < C_x\}} \tag{21}$$

If X is different but has the same cardinality as a bound of Y, then the constraint is satisfied:

$$(X: \text{ground}) \quad \frac{C_x = \#a_y, X \neq a_y}{\{X \neq Y\} \mapsto \{\}} \quad \frac{C_x = \#b_y, X \neq b_y}{\{X \neq Y\} \mapsto \{\}} \tag{22}$$

Symmetric inferences occur on instantiation of Y.

6.3.6 Disjointness

When two sets are told to be disjoint ($X \, \$ \, Y$, meaning $X \cap Y = \emptyset$), the sum of their cardinalities cannot exceed that of the union of their upper bounds (the implicit universe):

$$\frac{u = \#(b_x \cup b_y)}{\{tell(X \ \$ \ Y)\} \mapsto \{X \ \$ \ Y, C_x + C_y \le u\}} \tag{23}$$

Elements definitely in one set cannot belong to the other:

(X: changed glb)
$$\frac{b = b_y \setminus a_x}{\{X \ \$ \ Y\} \mapsto \{X \ \$ \ Y, Y \in [a_y, b]\}} \tag{24}$$

(Y: changed glb)
$$\frac{b = b_x \setminus a_y}{\{X \ \$ \ Y\} \mapsto \{X \ \$ \ Y, X \in [a_x, b]\}} \tag{25}$$

If the two sets are the same, they must have no elements (empty set):

(X or Y: bound)
$$\frac{}{\{X \ \$ \ Y, X = Y\} \mapsto \{X = \varnothing\}} \tag{26}$$

With the previous rules assured, it is enough that one set becomes ground to remove the constraint, since all its elements will have been removed from the other set:

(X or Y: bound)
$$\frac{ground(X) \vee ground(Y)}{\{X \ \$ \ Y\} \mapsto \{\}} \tag{27}$$

6.3.7 Set Inclusion

If Y contains X, then C_Y is greater (or equal) than C_X:

$$\frac{}{\{tell(X \subseteq Y)\} \mapsto \{X \subseteq Y, C_x \le C_y\}} \tag{28}$$

When the lower bound (*glb*) of X increases, the lower bound of Y may also increase; and when the upper bound (*lub*) of Y decreases, so might happen to X:

(X: changed glb)
$$\frac{a = a_x \cup a_y}{\{X \subseteq Y\} \mapsto \{X \subseteq Y, Y \in [a, b_y]\}} \tag{29}$$

(Y: changed lub)
$$\frac{b = b_x \cap b_y}{\{X \subseteq Y\} \mapsto \{X \subseteq Y, X \in [a_x, b]\}} \tag{30}$$

If b_X is contained in a_Y, or X is the same as Y, the constraint $X{\subseteq}Y$ is trivially satisfied, and can be eliminated from the store:

(X or Y: bound)
$$\frac{ground(X) \vee ground(Y), b_x \subseteq a_y}{\{X \subseteq Y\} \mapsto \{\}} \qquad \frac{}{\{X \subseteq Y, X = Y\} \mapsto \{X = Y\}} \tag{31}$$

6.3.8 Set Intersection

While for the set complement the universe must be given, for the intersection Z of sets X and Y, the universe can be considered as the union of the upper bounds ($\mathcal{U} = b_X \cup b_Y$), with cardinality u (as already used for set disjointness in rule number 23).

Whenever there are two set variables involved in a constraint, their interval domains, if depicted in a Venn diagram (Figure 6.4), define eight disjoint distinguishing sets of interest, each possibly empty. For instance, zone 6 of Figure 6.4 corresponds to the elements that are definitely part of both sets X and Y; in zone 4 are the elements that definitely belong to X but can never belong to Y; in zone 2 we have the elements that can belong to X or Y but are not yet definite elements of either. Constraint propagation over these constraints must take into account all set zones in addition to the two cardinality domains.

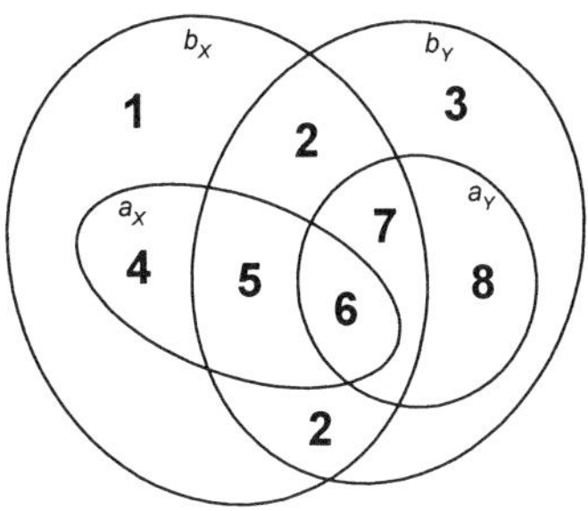

Figure 6.4. Two sets, X, Y, define 8 different zones

The following rule states that the intersection of two sets must be contained in both sets, and posts a special constraint on the cardinality of the set intersection:

$$\frac{}{\{tell(Z = X \cap Y)\} \mapsto \{Z = X \cap Y, tell(Z \subseteq X), tell(Z \subseteq Y), tell(C_z = X \otimes Y)\}} \quad (32)$$

The special cardinality constraint over sets ($C_z = X \otimes Y$) ensures that each possible value for C_z has a supporting cardinality pair in domains D_X and D_Y when the intersection is posted. Before formalising this operation, we first analyse what can the domain of cardinality C_z be. If we take possible cardinality values cx of D_X and cy of D_Y, and the sum of cx and cy exceeds u, there must be common elements to X and Y, and their intersection has at least $cx+cy-u$ elements. This 'worst' case is illustrated in Figure 6.5, where we 'push' all elements of X to the left and elements of Y to the right.

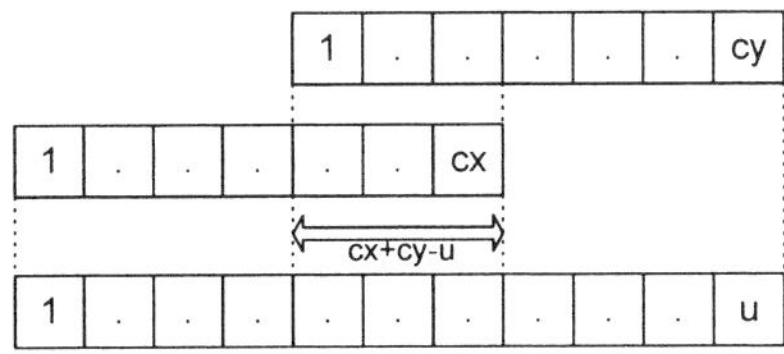

Figure 6.5. Minimum intersection cardinality

To reason about the upper bound, C_z can never exceed cx nor cy since Z is the sets intersection. The elements in a_X not in b_Y (i.e. $a_X \backslash b_Y$, corresponding to zone 4 of Figure 6.4) can safely be subtracted from cx since they are definitely not part of the intersection, but are counted in cx (so an upper bound can be $cx - \#a_X \backslash b_Y$). A similar reasoning may be done for Y, yielding another upper bound. A final upper bound can thus be considered the minimum of the two (i.e. $min(cx - \#(a_X \backslash b_Y), cy - \#(a_Y \backslash b_X))$).

Thus, for each pair cx and cy, an integer range for C_z is calculated, and the ranges for all such pairs are eventually merged. This can in fact be regarded as maintaining arc-consistency on the cardinality of X, Y and Z, when $Z = X \cap Y$. In fact, this arc-consistency is only enforced when the constraint is first told:

$$\frac{}{\{tell(C_z = X \otimes Y)\} \mapsto \{C_z \in \{n : \exists i \in D_x, j \in D_y, i + j - u \leq n \leq \min(i - \#(a_x \setminus b_y), j - \#(a_y \setminus b_x))\}\}} \tag{33}$$

The usefulness of this rule for the problems we address, can be illustrated with the differential diagnosis problem. Given two sets X and Y which can both be $\varnothing$ or $D=\{f,g\}$, let us consider their intersection (this is a typical case when two input bits are connected through an *and*-gate). While their set domain is the convex closure of the two bounds, their cardinality can only be 0 or 2. To find the cardinality domain of their intersection we examine cardinality pairs $\langle cx, cy \rangle = \langle 0,0 \rangle$, $\langle 0,2 \rangle$, $\langle 2,0 \rangle$ and $\langle 2,2 \rangle$. The three first pairs yield only value 0 as a possible intersection cardinality, since one set has no elements and acts as an upper bound. Pair $\langle 2,2 \rangle$ yields single value 2, since u is also 2 ($cx+cy-u=2+2-2=2$ as lower bound). Thus the final cardinality domain for the sets intersection is also $\{0,2\}$. If only interval reasoning were performed on cardinality, the result would be the full range $\{0,1,2\}$.

Rather than checking pairs of integers, it is equivalent and more efficient to check pairs of sub-ranges and their bounds when the constraint is posted. Nevertheless, since this arc consistency is very costly to maintain, it is only checked when the constraint is posted. Subsequently, only bounded arc consistency is maintained on the cardinality of the sets by the underlying finite domains constraint solver:

(*X*: changed glb or *Y*: changed lub) $\qquad$ $\dfrac{n = \#(a_x \setminus b_y)}{\{Z = X \cap Y\} \mapsto \{Z = X \cap Y, C_z \leq C_x - n\}}$ $\qquad$ (34)

(*X*: changed lub or *Y*: changed glb) $\qquad$ $\dfrac{n = \#(a_y \setminus b_x)}{\{Z = X \cap Y\} \mapsto \{Z = X \cap Y, C_z \leq C_y - n\}}$ $\qquad$ (35)

(*X* or *Y*: changed lub) $\qquad$ $\dfrac{n = \#b_x + \#b_y - \#(b_x \cup b_y)}{\{Z = X \cap Y\} \mapsto \{Z = X \cap Y, C_z \geq C_x + C_y - n\}}$ $\qquad$ (36)

A number of other inferences are performed regarding intersection. The lower bound of the set intersection is kept as the intersection of the lower bounds of the arguments (zone 6 of Figure 6.4):

(*X* or *Y*: changed glb) $\qquad$ $\dfrac{a = a_x \cap a_y}{\{Z = X \cap Y\} \mapsto \{Z = X \cap Y, Z \in [a,b_z]\}}$ $\qquad$ (37)

If both arguments are the same set, their intersection is that set (idempotence):

(*X* or *Y*: bound) $\qquad$ $\dfrac{}{\{Z = X \cap Y, X = Y\} \mapsto \{X = Y, tell(Z = X)\}}$ $\qquad$ (38)

If intersection Z is known to be the same set as one of its arguments, then the intersection constraint may be eliminated (as $Z \subseteq X$ and $Z \subseteq Y$):

(*X* or *Y*: bound) $\qquad$ $\dfrac{}{\{Z = X \cap Y, X = Z\} \mapsto \{X = Z\}}$ $\quad$ $\dfrac{}{\{Z = X \cap Y, Y = Z\} \mapsto \{Y = Z\}}$ $\qquad$ (39)

Conversely, if an argument contains the other, the intersection is the included set:

(*X* or *Y*: bound) $\qquad$ $\dfrac{ground(Y), b_x \subseteq a_y}{\{Z = X \cap Y\} \mapsto \{tell(Z = X)\}}$ $\quad$ $\dfrac{ground(X), b_y \subseteq a_x}{\{Z = X \cap Y\} \mapsto \{tell(Z = Y)\}}$ $\qquad$ (40)

Although inclusion could be inferred more generally, for efficiency reasons this rule is only checked when either one of the arguments is ground. These four simplification/simpagation rules [Frühwirth 1995] exploit the fact that the universe is the neutral element of the intersection. Here, the universe is the set argument containing the other.

All common elements to X and Y must be in Z. That is, X and Y must have no common elements outside Z. Hence, definite elements of argument X (Y) that are impossible in intersection Z must be removed from the other argument Y (X). This is a costly operation, so it is performed only on instantiation of variables:

$$(X \text{ or } Z: \text{ground}) \qquad \frac{b = b_y \setminus (a_x \setminus b_z)}{\{Z = X \cap Y\} \mapsto \{Z = X \cap Y, Y \in [a_y, b]\}} \qquad (41)$$

$$(Y \text{ or } Z: \text{ground}) \qquad \frac{b = b_x \setminus (a_y \setminus b_z)}{\{Z = X \cap Y\} \mapsto \{Z = X \cap Y, X \in [a_x, b]\}} \qquad (42)$$

In addition to the especial inferences exploiting the idempotence property and neutral element of set intersection (rules 38 to 40), which are triggered when some set is bound, let us now try to schematise inferences on variable bounds.

Globally, as to set bounds are concerned, we can make the inferences of Table 6.1, where some change in the set bound in the first column may have as effect an update of another set variable bound. Entries of the table thus express the update of the bound corresponding to its column due to some change of the bound corresponding to its line. Since lower bounds can only become larger, their updates are given in the form of a set of mandatory elements (i.e. another lower bound) — the final lower bound will thus be the union of the original one with this new one. Conversely, upper bounds become shorter and entry values correspond to another upper bound, which forces the intersection of the two upper bounds to obtain the final *lub*. Alternatively, upper bound updates may be given in the form –*(Set)* meaning that the elements of Set must be removed from the final *lub*.

Table 6.1. Set intersection: cause-effect rules on set bounds

Change \ Effect	a_x	b_x	a_y	b_y	a_z	b_z	
a_x			—	$-(a_x \setminus b_z)$	$a_x \cap a_y$	—	41, 37
b_x			—	—	—	b_x	$Z \subseteq X$
a_y	—	$-(a_y \setminus b_z)$			$a_x \cap a_y$	—	42, 37
b_y	—	—			—	b_y	$Z \subseteq Y$
a_z	a_z	—	a_z	—			$Z \subseteq X, Z \subseteq Y$
b_z	—	$-(a_y \setminus b_z)$	—	$-(a_x \setminus b_z)$			42, 41
	$Z \subseteq X$	42	$Z \subseteq Y$	41	37	$Z \subseteq X, Z \subseteq Y$	**Rules**

Each line also includes the inference rules that cover the updates due to its corresponding bound change. Similarly, each column includes the rules that cover the updates of its corresponding bound. To find the rule that covers a particular update given by some table entry, one just has to intersect the rules of its line with the rules of its column.

For example, from a change in a_x (i.e. inclusion of elements in X) one may infer that elements $(a_x \setminus b_z)$ must be removed from b_y (i.e. cannot be part of Y). The same happens when there is a change in b_z (i.e. removal of elements from Z). These relations are covered by rule number 41, which however is only triggered when X or Z is ground, for efficiency reasons.

As we can see from the table, each bound change affects some other bound, and each bound may be affected due to a change in another set variable bound. Among the rules that cover these updates are also the inclusion constraints (described in section 6.3.7) that are posted when the set intersection constraint is posted (rule number 32).

Rules 41 and 42 wait for some instantiation, while rule 37 and the inclusion constraint are immediately triggered with the set bound change.

Of course, bound changes also affect cardinalities, and cardinalities affect bounds in addition to being mutually dependent. These relations are handled by the constraints:

$$C_z \leq C_x - \#(a_x \backslash b_y)$$
$$C_z \leq C_y - \#(a_y \backslash b_x)$$
$$C_z \geq C_x + C_x - \#b_x + \#b_y - \#(b_x \cup b_y)$$

corresponding to rules 33 to 36 (also, cardinalities are nonnegative and limited by set bounds). After reaching arc consistency upon the posting of some set intersection, the constraint solver applies Bounded Arc Consistency (see section 1.3.1) on these cardinality constraints, which ensures that the bounds of cardinality domains are updated. In addition, such constraints are also updated due to changes in set bounds, which may further tighten the domains (e.g. when a_x increases or b_y decreases, $\#(a_x \backslash b_y)$ may increase and the upper bound of C_z given by $C_x - \#(a_x \backslash b_y)$ decreases, which further constrains C_z, and possibly C_x whose lower bound is then $C_z + \#(a_x \backslash b_y)$).

6.3.9 Set Union

For the union Z of sets X and Y, we can find many similarities with the intersection: the universe can also be considered as the union of the upper bounds ($\mathcal{U} = b_X \cup b_Y$), and u its cardinality.

The union of two sets, X and Y, must contain both sets, and its cardinality is not simply the sum of C_X and C_Y, since X and Y may have common elements. We know then that it can never exceed $C_X + C_Y$. In addition, the union constraint posts another special constraint on the resulting cardinality:

$$\frac{}{\{tell(Z = X \cup Y)\} \mapsto \{Z = X \cup Y, tell(X \subseteq Z), tell(Y \subseteq Z), C_z \leq C_x + C_y, tell(C_z = X \oplus Y)\}} \quad (43)$$

Before formalising the cardinality constraint ($C_z = X \oplus Y$), let us again first analyse what the domain of cardinality C_z may be. Given possible cardinalities cx of D_X and cy of D_Y, C_z can never exceed $cx + cy$ nor u, and the upper bound is thus $min(cx + cy, u)$. To find a lower bound, since Z is the sets union, it contains X and has at least the cx elements. To these we can safely add the elements in a_Y not in b_X (i.e. $a_Y \backslash b_X$, corresponding to zone 8 of Figure 6.4), since they are definitely part of the union and are not counted in cx. Similar reasoning is done for Y, yielding another lower bound, so the maximum of the two is the final lower bound given the $<cx,cy>$ pair (i.e. $max(cx + \#(ay \backslash bx), cy + \#(ax \backslash by))$). Thus, for each such pair, an integer range for C_z is calculated when the constraint is first told:

$$(44)$$
$$\frac{}{\{tell(C_z = X \oplus Y)\} \mapsto \{C_z \in \{n : \exists i \in D_x, j \in D_y, \max(i + \#(a_y \backslash b_x), j + \#(a_x \backslash b_y)) \leq n \leq \min(i + j, u)\}\}}$$

As an example, let us take two sets X and Y that can only be $\emptyset$ or $\{f,g,h,i\}$ (cardinality 0 or 4). To find the cardinality domain of their union we examine cardinality pairs $<0,0>$, $<0,4>$, $<4,0>$ and $<4,4>$. The three last pairs, yield only value 4 as a possible union cardinality, since at least one set has 4 elements, which is also the maximum possible value (u). Obviously, pair $<0,0>$ yields single value 0. Thus the final cardinality domain for the sets union is also $\{0,4\}$. This is a typical case for the above circuit models with two inputs passing through an or-gate. As with set intersection, if only interval reasoning were performed on the cardinality, the result would be the full range $0...4$ and no propagation would be achieved for the rest of the circuit.

Subsequently, bounded arc consistency is maintained on the cardinality of the sets:

$(X:$ changed glb or $Y:$ changed lub$)$ $$\frac{n = \#(a_x \backslash b_y)}{\{Z = X \cup Y\} \mapsto \{Z = X \cup Y, C_z \geq C_y + n\}} \quad (45)$$

$(X:$ changed lub or $Y:$ changed glb$)$ $$\frac{n = \#(a_y \backslash b_x)}{\{Z = X \cup Y\} \mapsto \{Z = X \cup Y, C_z \geq C_x + n\}} \quad (46)$$

$$(X \text{ or } Y: \text{ changed lub}) \qquad \frac{n = \#(b_x \cup b_y)}{\{Z = X \cup Y\} \mapsto \{Z = X \cup Y, C_z \leq n\}} \qquad (47)$$

As to set bounds, the upper bound of the set union is kept as the union of the upper bounds of the arguments:

$$(X \text{ or } Y: \text{ changed lub}) \qquad \frac{b = b_x \cup b_y}{\{Z = X \cup Y\} \mapsto \{Z = X \cup Y, Z \in [a_z, b]\}} \qquad (48)$$

If both arguments are the same set, their union is that set (idempotence):

$$(X \text{ or } Y: \text{ bound}) \qquad \frac{}{\{Z = X \cup Y, X = Y\} \mapsto \{X = Y, tell(Z = X)\}} \qquad (49)$$

If union Z is known to be the same set as one of its arguments, then the union constraint may be eliminated (as $X \subseteq Z$ and $Y \subseteq Z$):

$$(X \text{ or } Y: \text{ bound}) \qquad \frac{}{\{Z = X \cup Y, X = Z\} \mapsto \{X = Z\}} \quad \frac{}{\{Z = X \cup Y, Y = Z\} \mapsto \{Y = Z\}} \qquad (50)$$

Conversely, if an argument contains the other, the union is the container set:

$$(X \text{ or } Y: \text{ bound}) \qquad \frac{ground(Y), b_y \subseteq a_x}{\{Z = X \cup Y\} \mapsto \{tell(Z = X)\}} \quad \frac{ground(X), b_x \subseteq a_y}{\{Z = X \cup Y\} \mapsto \{tell(Z = Y)\}} \qquad (51)$$

These four rules exploit the fact that the universe (the set argument containing the other) is the absorbing element of the sets' union.

All elements of X and Y must be in Z. Thus, if Z has elements outside X (Y), then those elements must belong to Y (X). As a costly operation, it is performed only on instantiation of variables:

$$(X \text{ or } Z: \text{ ground}) \qquad \frac{a = a_y \cup (a_z \setminus b_x)}{\{Z = X \cup Y\} \mapsto \{Z = X \cup Y, Y \in [a, b_y]\}} \qquad (52)$$

$$(Y \text{ or } Z: \text{ ground}) \qquad \frac{a = a_x \cup (a_z \setminus b_y)}{\{Z = X \cup Y\} \mapsto \{Z = X \cup Y, X \in [a, b_x]\}} \qquad (53)$$

Inferences on set bounds and corresponding rules of set union constraint are depicted in Table 6.2, similarly to Table 6.1 for set intersection.

Table 6.2. Set union: cause-effect rules on set bounds

Change \ Effect	a_x	b_x	a_y	b_y	a_z	b_z	
a_x			—	—	a_x	—	$X \subseteq Z$
b_x			$a_z \setminus b_x$	—	—	$b_x \cup b_y$	52, 48
a_y	—	—			a_y	—	$Y \subseteq Z$
b_y	$a_z \setminus b_y$	—			—	$b_x \cup b_y$	53, 48
a_z	$a_z \setminus b_y$	—	$a_z \setminus b_x$	—			53, 52
b_z	—	b_z	—	b_z			$X \subseteq Z, Y \subseteq Z$
	53	$X \subseteq Z$	52	$Y \subseteq Z$	$X \subseteq Z, Y \subseteq Z$	48	**Rules**

Inclusion constraints are posted by rule number 43 (when union constraint is told).

Cardinality constraints are:

$$C_z \leq C_x + C_y$$
$$C_z \geq C_y + \#(a_x \setminus b_y)$$
$$C_z \geq C_x + \#(a_y \setminus b_x)$$
$$C_z \leq \#(b_x \cup b_y)$$

corresponding to rules 43 to 47.

6.3.10 Set Difference

For the difference Z of sets X and Y ($Z = X \setminus Y$), the universe is, as always, the union of the upper bounds ($\mathcal{U} = b_X \cup b_Y$), and u its cardinality.

The result Z of removing Y from X, must be contained in X and disjoint from Y, and its cardinality is not simply the difference of C_X and C_Y, since elements of Y may be absent from X. We know then that it is at least C_X - C_Y. In addition, the difference constraint posts another special constraint on the resulting cardinality ($C_Z = X \div Y$):

$$\frac{}{\{tell(Z = X \setminus Y)\} \mapsto \{Z = X \setminus Y, tell(Z \subseteq X), tell(Y \, \$ \, Z), C_z \geq C_x - C_y, tell(C_z = X \div Y)\}} \tag{54}$$

Let us then analyse what values can C_Z assume if we take possible cardinality values cx of D_X and cy of D_Y. It is at least cx - cy, as explained, since at most cy elements will be removed from X. Also, we can subtract from cx at most the number of common elements to both *lubs*, i.e. $\#(b_x \cap b_y)$. Hence, Z's cardinality is at least cx-$\#(b_x \cap b_y)$. Joining the two we have the lower bound $max(cx - cy, cx - \#(b_x \cap b_y))$.

As to the upper bound, since at least cy elements will not be a part of Z, we know that it contains at most u-cy elements. It can also never exceed cx. Furthermore, from cx we can safely subtract the definite common elements from X and Y, i.e. $\#(a_x \cap a_y)$. The upper bound is thus $min(cx - \#(a_x \cap a_y), u - cy)$ and we can formalise the posting of the cardinality constraint as follows:

$$\frac{}{\{tell(C_z = X \div Y) \mapsto \{C_z \in \{n : \exists i \in D_x, j \in D_y, max(i - j, i - \#(b_x \cap b_y)) \leq n \leq min(i - \#(a_x \cap a_y), u - j)\}\}} \tag{55}$$

We take again the two example sets X and Y that can only be $\varnothing$ or $\{f, g, h, i\}$ (cardinality 0 or 4). To find the cardinality domain of their difference $Z = X \setminus Y$, we examine cardinality pairs $<0,0>$, $<0,4>, <4,0>$ and $<4,4>$. When cy=4, the resulting cardinality must be 0 (upper bound given by u - cy) since u is also 4. If cx=0, the resulting cardinality must also be 0 (bounded by cx - $\#(a_x \cap a_y)$). The remaining pair, $<4,0>$, yields only value 4 as a possible difference cardinality, since no elements are removed from set X with 4 elements (#Z is lower bounded by cx - cy). The final cardinality domain for the sets difference is thus also $\{0,4\}$.

For another example with different sets and cardinalities, let us take $X \in [\{\}, \{e_1, e_2, \ldots e_{20}\}]$ 10, and $Y \in [\{\}, \{e_{20}, e_{21}, \ldots e_{30}\}]$. In this case we may conclude that the cardinality of $Z = X \setminus Y$ has at least 9 elements due to the lower bound cx - $\#(b_x \cap b_y) = 10 - \#\{e_{20}\} = 10$-$1 = 9$. The domain of the cardinality of Z is then constrained to $\{9,10\}$ using just this rule.

Let us now consider two sets X, Y with domain $[\{e_1, e_2, \ldots e_5\}, \{e_1, e_2, \ldots e_{20}\}]$ 6. Their difference has at most 1 element due to the bound cx - $\#(a_x \cap a_y)$. This rule thus allows us to constrain the cardinality of difference Z to the simple domain $\{0,1\}$ from start.

Subsequently, as with the intersection and union operations, bounded arc consistency is applied on the cardinality of the sets:

$$(X \text{ or } Y \text{: changed glb}) \qquad \frac{n = \#(a_x \cap a_y)}{\{Z = X \setminus Y\} \mapsto \{Z = X \setminus Y, C_z \leq C_x - n\}} \qquad (56)$$

$$(X \text{ or } Y \text{: changed lub}) \qquad \frac{n = \#(b_x \cup b_y)}{\{Z = X \setminus Y\} \mapsto \{Z = X \setminus Y, C_z \leq n - C_y\}} \qquad (57)$$

Regarding set bounds, definite elements of X that cannot be removed (not part of Y) must be included in Z:

$$(X \text{: ch. glb or } Y \text{: ch. lub}) \qquad \frac{a = a_z \cup (a_x \setminus b_y)}{\{Z = X \setminus Y\} \mapsto \{Z = X \setminus Y, Z \in [a, b_z]\}} \qquad (58)$$

Conversely, definite elements of X that cannot be part of Z must be included in Y, so that they are removed:

$$(X \text{: ch. glb or } Z \text{: ch. lub}) \qquad \frac{a = a_y \cup (a_x \setminus b_z)}{\{Z = X \setminus Y\} \mapsto \{Z = X \setminus Y, Y \in [a, b_y]\}} \qquad (59)$$

If both arguments are the same set, their difference is empty:

$$(X \text{ or } Y \text{: bound}) \qquad \frac{}{\{Z = X \setminus Y, X = Y\} \mapsto \{X = Y, Z = \varnothing)\}} \qquad (60)$$

If $X=Z$, then we can remove the difference constraint, since we have already constrained Y and Z to be disjoint:

$$(X \text{ or } Y \text{: bound}) \qquad \frac{}{\{Z = X \setminus Y, X = Z\} \mapsto \{X = Z\}} \qquad (61)$$

If arguments are disjoint, then the difference is the first set:

$$(X \text{ or } Y \text{: bound}) \qquad \frac{(ground(X) \vee ground(Y)), X \cap Y = \varnothing}{\{Z = X \setminus Y\} \mapsto \{Z = X\}} \qquad (62)$$

If both arguments are ground, we can remove the constraint:

$$(X \text{ or } Y \text{: bound}) \qquad \frac{ground(X), ground(Y), z = X \setminus Y}{\{Z = X \setminus Y\} \mapsto \{Z = z\}} \qquad (63)$$

Elements of X must be present either in Y or in Z, since if such an element is not in Y then it is not removed and is forcefully in difference Z. Consequently; the universe of Y and Z limits X. As a costly operation, it is performed only once, on instantiation of Z:

$$(Z \text{: ground}) \qquad \frac{b = (b_x \cap (b_y \cup b_z))}{\{Z = X \setminus Y\} \mapsto \{Z = X \setminus Y, X \in [a_x, b]\}} \qquad (64)$$

Inferences on set bounds and corresponding rules of set difference constraint are depicted in Table 6.3, similarly to the previous two sections.

Inclusion and disjointness constraints are posted by rule number 54 (when set difference constraint is told).

Constraints involving cardinalities are:

$$C_z \geq C_x - C_y$$
$$C_z \leq C_x - \#(a_x \cap a_y)$$
$$C_z \leq \#(b_x \cup b_y) - C_y$$

corresponding to rules 54 to 57.

Table 6.3. Set difference: cause-effect rules on set bounds

Change \ Effect	a_x	b_x	a_y	b_y	a_z	b_z	Rules
a_x			$a_x \backslash b_z$	—	$a_x \backslash b_y$	—	59, 58
b_x			—	—	—	b_x	$Z \subseteq X$
a_y	—	—			—	$-(a_y)$	$Y \$ Z$
b_y	—	$b_y \cup b_z$			$a_x \backslash b_y$	—	64, 58
a_z	a_z	—	—	$-(a_z)$			$Z \subseteq X, Y \$ Z$
b_z	—	$b_y \cup b_z$	$a_x \backslash b_z$	—			64, 59
Rules	$Z \subseteq X$	64	59	$Y \$ Z$	58	$Z \subseteq X, Y \$ Z$	**Rules**

6.4 Implementation

We used ECLiPSe with attributed variables to implement *Cardinal*, the set constraint solver with cardinality inferences based on the above rules. The attributes of a set variable are its domain and its cardinality together with lists of suspended goals, since we used the underlying predicate suspension handling mechanism. Note that ECLiPSe provides waking conditions such as a change in some domain, but the waking constraint does not know what exactly has changed (it only may know that it has changed somehow), which is a possible source of inefficiency. For instance, with constraint $X \subseteq Y$, if element a is inserted in X, we know immediately that a must also be inserted in Y. Unfortunately, with ECLiPSe mechanisms we have to wake the constraint due to a change in X's *glb*, and then the constraint must include the whole X's *glb* in Y, since it does not know what are the new elements. This can be overcome with reactive changes as in [Zhou 2000]. Notwithstanding such limitations, good results were still achieved.

The cardinality of a set is an integer variable to be handled by the ECLiPSe finite domain library. To represent the domain of set S as a set interval we need its bounds a_s and b_s. Since $a_s \subseteq b_s$, it is enough to store a_s as the definite elements of S, and the difference $b_s \backslash a_s$ as the possible extra elements of S (both implemented as sorted lists). For efficiency reasons, the sizes of its two bounds are also stored.

Set complement constraints take as arguments the universe and the input and output set variables. In general, constraints perform all the possible inferences when posted (interval-consistency on the sets; arc-consistency on their cardinalities), while their subsequent maintenance only ensures arc-consistency on their bounds. The rationale for this is that it is worth spending more time trying to reduce domains, only if this effort is not done too often.

6.4.1 Set Labelling

Cardinal, as the majority of constraint solvers, is not complete, which means that even when constraint propagation is successful, CSP variables must still be instantiated in order to prove that a solution is possible (or not). Since set variables are represented by set intervals, we can split the search space in two (similarly to the usual way of handling intervals over reals) by a disjunction on the membership of a set's possible extra elements. I.e. $x \in b_s \backslash a_s$ either belongs or not to set S. Hence, at each such try (disjunction solving), the set domain is restricted either by adding x to a_s or by removing it from b_s. This allows making a more active use of constraints during the search phase, avoiding instantiating the set directly to some element in the domain. Such direct instantiation of set variable S to a particular ground set s can be hard to succeed for large domains (usually the case, with set intervals), and if it fails, then only another constraint, $S \neq s$, is derived, which hardly will restrict S (or any other variable) domain. This naïve labelling, generally, only succeeds after many failures. *Conjunto* implements set labelling as a recursive refine procedure over disjunctions of the form (x in S ; x notin S) for possible elements x of set S. This means that inclusion of x is always tried first. However, we realised that often a labelling strategy of first trying exclusion is more effective. Hence, in *Cardinal*, we implemented set labelling with an extra parameter (*up* or *down*) that indicates what choice to try first. When given value '*up*', a set is labelled as in *Conjunto*, while for value '*down*', choices are handled as (x notin S ; x in S), which is actually the default in *Cardinal*.

6.5 Results

Gates are currently implemented based on two basic set operations alone; namely, complement (*not*-gate) and binary intersection (*and*-gate). Exclusive set union (*xor*-gate) is not currently implemented as a basic operation, hence it is defined as "$S_1 \oplus S_2$" $\equiv (S_1 \cap \overline{S_2}) \cup (\overline{S_1} \cap S_2) = \overline{(\overline{S_1 \cap \overline{S_2}})} \cap \overline{(\overline{\overline{S_1} \cap S_2})}$.

Our labelling strategy for circuit problems finds for the relevant output bits and S-buffers, the inputs they depend on. These are then labelled by assigning values (0 or u) to their cardinality. If successful, the rest of the circuit input is labelled.

To assess the advantages of cardinality inferences, we took off inferences from *Cardinal* that are not implemented in *Conjunto* (i.e. rewrite rules regarding cardinality and other especial inferences, numbered 11 of set complement; 28 of set inclusion; 32 to 36 and 38 to 42 of set intersection). As mentioned in the introduction of this chapter, this simpler version is referred to as "Conjunto". We then tried to solve the same problems for the standard ISCAS digital circuits benchmarks [ISCAS 1985] using *Cardinal* and "Conjunto". From a set of differentiation benchmarks created over these circuits [Azevedo and Barahona 1999], we randomly picked pairs of diagnoses to differentiate. The results are shown in Table 6.4 (times reported in seconds on a Pentium III, 500 Mhz).

Table 6.4. Experimental Results

circuit	Diag1	Diag2	Diff.	"Conjunto"	Cardinal	Speed-up
c432	380gat/0	415gat/1	X	8.1	0.4	20.3
	431gat/0	428gat/1	√	39.4	1.3	30.3
	431gat/0	419gat/0	√	37.9	1.4	27.1
	428gat/1	419gat/0	X	24.0	1.0	24.0
c1908	1541/1	1538/0	√	24.2	3.2	7.6
	860/1	72/1	√	156.3	1.3	120.2
	72/1	71/0	X	194.9	1.0	194.9
c3540	855/0	707/1	√	4.1	6.1	0.7
	955/0	954/0	X	3482.8	2.4	1451.2
	855/0	707/1	√	1.8	3.2	0.6
	403/0	3544/1	√	352.1	2.9	121.4
c6288	5671gat/0	5537gat/1	√	> 86400	11.0	> 7854.5
	6288gat/1	6285gat/0	√	> 3600	8.9	> 404.5
	813gat/0	6123gat/0	√	> 3600	8.9	> 404.5

This table, with results for 4 ISCAS circuits, indicate the time that *Cardinal* and "Conjunto" needed to find a differentiating test pattern between *Diag1* and *Diag2* (marked as √) or to prove it is impossible (i.e. the faults are indistinguishable, shown as X). For example, the first line reports that the differentiation of gate *380gat* stuck-at-0 from gate *415gat* stuck-at-1, in circuit *c432*, took 8.1 seconds in "Conjunto" and 0.4 seconds in *Cardinal*.

Globally, it can be stated that *Cardinal* showed a speed-up of two orders of magnitude compared to "Conjunto" on this set of problems (and others we tried) although, as expected, the improvement was not uniform over all the tests.

While for circuit *c432* Cardinal showed a consistent speed-up around 25, for larger circuits the variation can be quite large. In circuits *c1908* and *c3540*, the speed-up ranges from 0.6 to 1451.2, *Cardinal* being more efficient in harder problems (specially those where there is no differentiating pattern between the two diagnoses). For the two instances in circuit *c3540* where there was an easy solution for "Conjunto", *Cardinal* was slower due to the extra inferences performed, and the times thus reflect this overhead. The extra computing effort may be largely compensated, as tests in *c6288* show, where *Cardinal* easily found a solution, whereas "Conjunto" had to be aborted in all three tests after one hour (one particular test was even kept running for one day of unsuccessful processing). Of course, the speed-up can be arbitrarily large as long as not enough propagation was achieved and we start labelling variables, since the execution time is exponential on the number of these variables.

Due to all the especial inferences and list processing, we expected *Cardinal* to experience problems with larger circuits or in problems with many diagnoses. Also, since the general feeling is that, in practice, it is very costly to perform all the desired inferences over sets and their cardinalities, we tried to create another version with n-ary gates but with fewer inferences, which produced results that were midway between "Conjunto" and *Cardinal*. The fact is that *Cardinal* still managed to efficiently solve problems for the largest of the benchmark circuits (*c7552*), so no improvements were obtained by reducing inferences.

6.6 Other Applications

In this section we present other possible applications for *Cardinal*, together with experimental results and comparisons to other approaches. More complete results on the differentiation problem are also presented at the end of the section. Another general application, set covering, is discussed in section 6.7.3 when analysing Cardinal extensions.

6.6.1 Steiner Triples

The ternary Steiner problem of order n consists of finding a set of $n.(n\text{-}1)/6$ triples of distinct integer elements in $\{1,\ldots,n\}$ such that any two triples have at most one common element. It is a hypergraph problem coming from combinatorial mathematics [Lueneburg 1989] where n modulo 6 has to be equal to 1 or 3 [Lindner and Rosa 1980], and which has recently been addressed in computer science [Beldiceanu 1990, Gervet 1997]. One possible solution for n=7 (Steiner 7) is $\{\{1, 2, 3\}, \{1, 4, 5\}, \{1, 6, 7\}, \{2, 4, 6\}, \{2, 5, 7\}, \{3, 4, 7\}, \{3, 5, 6\}\}$. The solution contains $7*(7\text{-}1)/6 = 7$ triples.

A CLP(*Sets*) approach easily models this problem by representing each triple as a set variable with cardinality 3, and constraining the cardinality of each intersection of a pair of triples to be not greater than 1. In contrast, an integer domain CSP model would require the triple of variables, and far more constraints, in addition to not being so declarative. Furthermore, since set elements are not ordered, much symmetry that could occur in the integer approach is naturally eliminated with set variables.

Thus, for Steiner n, we declare $n.(n\text{-}1)/6$ set variables with domain $[\{\},\{1,\ldots,n\}]_3$ and, for each pair S_1, S_2 of such variables we impose $\#(S_1 \cap S_2) \leq 1$. Table 6.5 presents the results (seconds to find a solution) obtained with this model for *Cardinal* and our version of *Conjunto* in a Pentium 166. For Steiner 7 and 9 the number of triples are, respectively, 7 and 12.

Table 6.5. Steiner triples results

Steiner	Triples	"Conjunto"	Cardinal
7	7	0.72	0.43
9	12	356.35	213.49

Note that the complexity grows exponentially with n and Beldiceanu [1990], using CHIP and global constraints, does not present results for $n=9$.

An improvement for this particular problem, which we did not implement but may drastically reduce computation for large n's [Gervet 1997], consists in constraining each element to belong to at most $(n-1)/2$ triples. This can be done during labelling by a simple 'occurs-check' when adding an element to a set. This check is justified since there can be no two equal elements in two triples, i.e. no equal pairs. With the $\{1,...,n\}$ domain, there are at most $n-1$ distinct pairs containing element i and since a triple containing i forcefully contains 2 such pairs, we reach the maximum of $(n-1)/2$ triples containing i.

6.6.2 Golfers

The golfers tournament (or social golfer) problem was proposed in 1998 by Warwick Harvey (described on the web in [Gent and Walsh 1999]) after a question posted to sci.op-research. It is a generalisation of the problem of constructing a round-robin tournament schedule, where the number of players in a "game" is more than two.

A particular goal, for 32 players in 8 groups of 4 each week, is to set up the foursomes so that each person only golfs with the same person once, and that the number of weeks to play is maximised. The optimal solution for 32 golfers is not yet known, but since a golfer plays with 3 new people each week, the schedule cannot exceed 10 weeks.

Cardinal can find a solution for 9 weeks by declaring 9 times 8 set variables. Let S_{ij} refer to the set variable corresponding to week i ($i \in \{1..9\}$) and foursome j ($j \in \{1..8\}$). The constraints are thus:

$$S_{i,j}:[\{\},\{1,...,32\}]_4$$

$$\forall i, \bigcup_j S_{i,j} = \{1,...,32\}$$

$$\forall i_1 \neq i_2, \forall j_1 \neq j_2, \#(S_{i1,j1} \cap S_{i2,j2}) \leq 1$$

Each variable is a foursome of golfers and the union of the 8 foursomes of a week is the whole set of 32 golfers. The constraint that two players cannot meet twice is assured by picking each pair of foursomes of different weeks, and forcing the cardinality of their intersection to be less or equal than 1.

On a Pentium II / 450Mhz *Cardinal* under Linux found a solution in 38 seconds while *Conjunto* required 44 seconds. Note that this experiment was conducted with the real *Conjunto* solver, not our version of it.

6.6.3 Warehouse

The warehouse problem is an optimisation problem consisting in deciding which warehouses to build from a set of known locations, so that a given set of customers is served with the minimum cost. There is a cost per customer being delivered from a specific warehouse, whereas the cost of building a new warehouse is constant and unique: 50000.

A particular instance of this problem, with 20 customers and 19 warehouses, is shown in Table 6.6, with the optimum solution stressed (built warehouses in bold, deliveries shaded).

Table 6.6. Delivery costs of warehouse problem

	1	2	3	4	5	6	7	8	9	10
Roissy	68948	15724	24300	4852	40950	66330	39698	45895	5519	53433
Orly	68948	8634	24300	4852	40950	66330	39698	23975	4387	53433
Lille	68948	17850	12600	4852	78390	66330	39698	45895	9481	53433
Nancy	68948	17850	24300	2817	31590	66330	39698	45895	9481	53433
Lyon	35101	23520	60300	4852	16380	34650	15998	23975	9481	21533
Rouen	68948	16433	24300	6104	40950	66330	39698	45895	4387	53433
Toulouse	24524	46200	60300	10486	40950	13860	14813	18495	5519	21533
Montpellier	24524	46200	60300	10486	31590	24750	8295	23975	9481	19938
Bordeaux	35101	23520	60300	10486	40950	26730	20738	9590	4387	27913
Nantes	68948	17850	60300	10486	78390	34650	39698	18495	2547	53433
Marseille	26639	46200	60300	10486	31590	26730	14813	23975	9481	11165
Nice	35101	46200	60300	10486	31590	34650	15998	45895	9481	19938
Dijon	68948	17850	31500	4852	29250	34650	20738	45895	5519	27913
Rennes	68948	17850	31500	10486	78390	66330	39698	23975	4104	53433
Tours	68948	17850	31500	6104	40950	34650	20738	18495	4387	53433
Orleans	68948	15724	24300	4852	31590	34650	20738	23975	4387	53433
Chalons	68948	16433	24300	4539	40950	66330	39698	45895	5519	53433
Clermont	26639	17850	60300	6104	29250	26730	15998	18495	5519	21533
Annecy	35101	23520	60300	4852	28080	34650	15998	45895	9481	21533

	11	12	13	14	15	16	17	18	19	20
Roissy	47235	13125	138176	106993	55408	55786	16847	27780	4278	9672
Orly	47235	13125	159783	123723	47915	93864	16847	24308	3983	9005
Lille	47235	14175	181390	140454	62900	106556	19125	31253	4573	12340
Nancy	47235	18375	181390	140454	62900	106556	19125	31253	4573	12340
Lyon	19035	18375	239008	185069	82880	140403	25200	31253	4573	9672
Rouen	47235	7350	166985	129300	57905	98095	17607	28938	4573	9672
Toulouse	24675	35175	469480	363528	162800	275792	49500	77553	5753	12340
Montpellier	19035	35175	469480	363528	162800	275792	49500	77553	5753	12340
Bordeaux	47235	35175	239008	185069	82880	140403	25200	40513	5753	9672
Nantes	47235	14175	239008	140454	82880	140403	19125	40513	4573	9672
Marseille	17625	35175	469480	363528	162800	275792	49500	77553	9883	22345
Nice	9870	35175	469480	363528	162800	275792	49500	77553	9883	22345
Dijon	24675	18375	181390	140454	62900	106556	19125	31253	4573	9672
Rennes	47235	14175	181390	140454	62900	106556	19125	31253	4573	9672
Tours	47235	14175	181390	140454	62900	106556	19125	31253	4278	8671
Orleans	47235	14175	166985	129300	57905	98095	17607	28938	2655	8671
Chalons	47235	14175	166985	129300	57905	98095	19125	28938	4573	9672
Clermont	24675	18375	239008	185069	62900	140403	25200	31253	4573	9672
Annecy	19035	35175	239008	185069	82880	140403	25200	40513	5753	12340

An ILP model of this problem uses nineteen 0-1 variables W_i to express whether warehouse i ($i \in \{1..19\}$) is built or not. Twenty customers variables C_j take $\{1..19\}$ as integer domain, meaning that customer j ($j \in \{1..20\}$) is served by warehouse C_j with cost given by *cost(j,Cj)* as the respective value of the costs table. The cost we want to minimise is then

$$Cost = 50000 * \textstyle\sum_i W_i + \sum_j cost(j, C_j)$$

A set model uses a single set variable *Ws* for the built warehouses, with an empty *glb* and a *lub* consisting of all warehouses $\{1..19\}$. Customer deliveries may also be modelled as the 20 integer variables C_j of ILP. The number of warehouses in the cost function is now simply #*Ws* instead of a sum of 19 variables.

Implementation of these models in ECLiPSe use built-in constraints *element/3* and *sumlist/2* to, respectively, retrieve delivery costs and sum them. Since we want to minimise costs, we start labelling by trying to eliminate warehouses (assign 0 to W_i variables, or reduce *lub* of *Ws*). As soon as we decide not to build warehouse i, it must be removed from the domains of C_j. This is assured, for each customer j, by constraint *occurrences*$(i,C_j,0)$ in ILP, or by $C_j::lub(Ws)$ in the sets model. Delivery costs are labelled by picking the minimum value of the table column for each customer. The minimum cost is 730567 with warehouses built in Roissy and Montpellier. The solution given by the sets model is $Ws=\{1,8\}$ (or {Roissy, Montpellier}), which is more natural than the ILP solution for 19 variables (1000000100000000000). The program is also more declarative in CLP(*Sets*). We tested this problem for ILP, *Cardinal*, "Conjunto" and the real *Conjunto* on a Pentium III / 500Mhz, all of them reaching and proving the optimum in about 55 hundredths of a second. We first notice that set constraint programming achieved the same result as ILP. Comparing the different set solvers, for this problem where cardinality was not particularly important, *Cardinal* behaved as "Conjunto", thus showing no overhead for cardinality inferences. "Conjunto" also behaved as the real *Conjunto*. (For equivalence, we implemented the labelling strategy, cf. section 6.4.1, of refining down the warehouses in the *Conjunto* model, since it is not built-in.)

6.6.4 Differential Diagnosis

We have already shown in section 6.5 some results for the differentiation problem. In Table 6.7 we present more complete results for different set libraries over ECLiPSe with Linux on a Pentium II / 450Mhz, obtained in the year 2000 during a stay in IC-Parc in cooperation with Joachim Schimpf, Carmen Gervet and Mark Wallace. The real *Conjunto* is tested together with *Cardinal* and *fd_sets* (a new optimised set library of ECLiPSe for integer sets, by the ECLiPSe team and Neng-Fa Zhou).

Table 6.7. Differentiation results over different set libraries

circuit	Diag1	Diag2	Diff.	Conjunto	fd_sets	Cardinal
c432	380gat/0	415gat/1	X	12.0	7.3	0.4
	431gat/0	428gat/1	√	7.0	4.4	0.4
	431gat/0	419gat/1	√	7.0	4.4	0.4
	428gat/1	419gat/1	√	9.7	6.1	0.5
	431gat/0	419gat/0	√	6.9	4.4	0.4
	428gat/1	419gat/0	X	0.8	6.3	0.3
c499	od2/0	id2/0	√	0.2	6.1	2.5
	od2/0, od19/0	id2/0, od19/0	√	0.3	6.1	2.5
c880	389gat/1	291gat/1	X	1.6	11.4	0.5
	422gat/0, 850gat/0	422gat/0, 840gat/0	X	0.1	1.2	0.5
c1355	1258gat/1	1339gat/0	√	688.2	189.9	1.0
	162gat/0	1274gat/1	√	968.0	0.7	1.1
	1347gat/0	1315gat/0	√	686.0	391.5	0.8
c1908	1541/1	1538/0	√	14.3	11.2	1.3
	860/1	72/1	√	251.9	185.7	1.3
	72/1	71/0	X	315.6	232.8	1.1
c2670	96/1	221/0	√	0.5	1.2	1.7
	217/0	216/1	X	37.3	22.3	1.4
	162/1	1467/0	√	1.1	1.5	1.8
	236/0	120/1	√	0.5	1.2	1.7
c3540	855/0	707/1	√	0.5	1.4	2.2
	955/0	954/0	X	5579.1	3366.0	1.9
	403/0	3544/1	√	na	335.3	2.3
c5315	91/0	742/0	√	6.6	5.9	3.5
	651/1	649/1	√	7.7	6.6	3.5
	649/1	1598/1	X	296.4	188.6	2.8
c6288	5671gat/0	5537gat/1	√	na	na	4.7
	6288gat/1	6285gat/0	√	na	na	4.6
	813gat/0	6123gat/0	√	na	na	4.6
c7552	5222/1	5221/1	X	16.9	150.9	5.6
	4772/1	330/1	√	14.3	12.2	4.8

Again we are trying to differentiate two diagnoses (*Diag1* and *Diag2*) over the ISCAS circuits. Such a pair of diagnoses (each consisting of one or two stuck gates) can be differentiable (marked with √) or not (marked with X). Results are in seconds and '*na*' values mean non-available results (i.e. aborted execution after hours of processing). As explained in the introduction of this chapter, *Conjunto* results are not very reliable since this solver was still 'buggy' at the time (shaded values represent tests where an incorrect solution was detected). Nevertheless, we think that these values provide us with an overall idea of the mean execution time of *Conjunto* over these problems.

As can be seen in the table, *Cardinal* is always able to find the solution in 0.4 to 5.6 seconds, taking 1 minute to solve the whole set of 31 tests, while even *fd_sets* could not solve any differentiation for *c6288*, and for other circuits it can take 2 or 5 minutes or even an hour to solve a single test.

To compare the performance of our set model with the one using an 8-valued logic presented in Chapter 4, we ran the same set of differentiation tests, this time on a Pentium 4, 1.7 GHz, 512 Mb RAM, using ECLiPSe Prolog 5.1 (Tcl/Tk version) under Windows 2000.

Both versions use the *ITBS-Path* heuristic described in section 4.9, with the difference that when using sets it is the cardinality that is labelled (with value 1) for each signal in the chosen path. Although generally faster, the 8-valued logic model required about 74 seconds to solve all tests, while *Cardinal* required only 23 seconds. This was due to a single test in circuit *c6288*, which took one minute solving, while *Cardinal* never needed more than two seconds for each test.

For a more exhaustive comparison, we picked the smallest and largest ISCAS circuits, namely *c432* and *c7552*, and executed the respective differentiation benchmarks (partitioning diagnoses into sets of equivalence classes) as described in section 4.7, to obtain the results of Table 6.8, where total time for each benchmark is shown in seconds for both solvers.

Table 6.8. Differentiation results with different solvers

	N	NC	ND	8V	Sets
b432_a_1	4	1	3	0.09	0.33
b432_a_2	6	5	11	44.00	3.78
b432_b_1a	3	2	2	0.08	0.30
b432_b_1b	2	2	1	0.06	0.17
b432_b_2	4	4	6	0.30	3.02
b432_c_1	2	2	1	0.03	0.19
b7552_a_1a	13	9	43	55.38	101.02
b7552_a_1b	13	9	43	53.86	112.63
b7552_b_1	13	12	73	223.75	179.47
b7552_c_1a	16	6	25	22.87	49.24
b7552_c_1b	14	7	29	30.12	61.95

Again, *N* denotes the number of initial diagnoses; *NC* is the number of computed classes and *ND* is the number of differentiation tests required to obtain the final partition. The trend for the 8-valued logic solver (denoted as 8V) to be faster than using sets is confirmed in general but, interestingly, harder instances take longer time to solve. In fact, for *c7552*, 8V solves four benchmarks about twice as fast but requires much more time for *b7552_b_1*, for which *Cardinal* is faster. Also, with *c432*, 8V required 44 seconds to solve *b432_a_2*, while *Cardinal* solved it in less than 4 seconds.

Thus, it seems that the extra complexity of set solving, although heavier for easier problems, produces more consistent results by being less dependent on phase transitions [Cheeseman *et al.* 1991].

6.7 Cardinal Extensions

In the previous sections we presented a new set constraint solver, *Cardinal,* that makes a number of special inferences over the sets cardinality to outperform other similar solvers. In fact, set constraints are a very natural and concise way to express problems such as set covering, partitioning or bin-packing. Furthermore, actively considering set functions such as the cardinality, allows to efficiently solve these problems and to express many optimisation goals. In this section we extend the constraint solving of *Cardinal* to set variables with attached set functions with special inferences over them [Azevedo and Barahona 2000b] and we exemplify its use with some applications. Hence, this section focuses on combinatorial search problems modelled in *Cardinal.* In particular, after diagnostic problems were considered where cardinality played a special role, we address set covering problems where the sets union function is used to make additional inferences thus pruning the search space. We then compare experimental results with other approaches to attest the expressiveness and efficiency of *Cardinal.* Possible extensions of *Cardinal* are also discussed for other applications such as timetabling and other scheduling problems.

6.7.1 Sets Union

Inference capabilities over set functions, such as the cardinality, are particularly important when solving set problems where they play a special role. Other possible graded functions are the minimum and maximum of sets of numbers. Constraints over such functions considering integer ranges for these minima or maxima could also be very useful since problems with such constraints are very common. Of course, these functions may be applied to sets of elements of arbitrary type, not just numbers, as long as a valid ordering function is associated (like Prolog's lexicographic @< / 2 for 2 arbitrary terms). Therefore, there is no reason to consider only functions mapping sets to integers. An interesting function is the sets union mapping sets of sets to sets. This function maps set $\{\{a,b\},\{a,f\},\{b,c\},\{g\}\}$ into set $\{a,b,c,f,g\}$ and satisfies (for two sets of sets s_1, s_2) $s_1 \subseteq s_2 \Rightarrow \bigcup s_1 \subseteq \bigcup s_2$.

6.7.2 Generalisation to Sets Functions

In extended *Cardinal,* a set variable X may be represented by $[in_X + poss_X]\text{-}Functions(X)$ where in_X is its greatest lower bound (i.e. the elements known to belong to X), $poss_X$ the set of extra elements still possible in X ($poss_X$ and in_X are disjoint and their union constitutes the lowest upper bound of X, i.e. the elements not excluded from X), and $Functions(X)$ is a set of functions of X. Each element of $Functions(X)$ is a pair $f{:}V$ where f is the function identifier, and V a domain variable representing its value $f(X)$. The currently possible functions are the cardinality ($\#$) and the union ($\cup$) functions.

For example, $[\{a,g\} + \{b,c,x,y\}]\text{-}\{\#{:}C\}$ where C is a finite domain variable with domain $\{3,5\}$, represents a set variable with 3 or 5 elements, being two of them a and g and the rest coming from only $\{b,c,x,y\}$.

The cardinality function is an integer variable, whereas the union function can be a normal *Cardinal* set variable with associated functions itself.

6.7.3 Set Covering

Set covering is an optimisation or satisfaction problem well studied in Operations Research (OR) and that often occurs in real life, as was the case with the minimisation problem in circuits (section 5.2.1).

Let us consider a simple satisfaction example in the usual 0-1 Integer Linear Programming (ILP) approach:

$$\begin{bmatrix} 1 & 0 & 0 & 1 \\ 1 & 1 & 0 & 0 \\ 0 & 0 & 1 & 1 \\ 0 & 1 & 0 & 1 \\ 1 & 0 & 1 & 0 \end{bmatrix} \cdot \begin{bmatrix} x_1 \\ x_2 \\ x_3 \\ x_4 \end{bmatrix} \geq \begin{bmatrix} 1 \\ 1 \\ 1 \\ 1 \\ 1 \end{bmatrix} \qquad \text{and} \qquad x_1 + x_2 + x_3 + x_4 = 2$$

where each of the 4 x_i is a 0-1 variable.

Intuitively, the goal is to find 2 columns in the 5x4 matrix, where we can find at least one value 1 on each line. The solution is then $x_1 = 1$ and $x_4 = 1$.

Given a set-covering problem, a set-based model should be quite natural. What we want to find is indeed simply a set of two columns so that all lines are "covered". One way to represent this is:

$Cols \subseteq \{1,2,3,4\}, \quad \#Cols=2,$
$\{1,4\} \cap Cols \neq \varnothing$
$\{1,2\} \cap Cols \neq \varnothing$
$\{3,4\} \cap Cols \neq \varnothing$
$\{2,4\} \cap Cols \neq \varnothing$
$\{1,3\} \cap Cols \neq \varnothing$

Now the solution is $Cols = \{1,4\}$.

But *Cardinal* has yet another modelling solution to this problem by means of the implemented union function of a set variable. Although the cardinality function is always present with a finite domain variable, the union function is optional since it only applies to sets of sets.

For the previous example with the union function, we can simply declare a set variable S as:

$$S :: [\{\} + \{\{1,2,5\},\{2,4\},\{3,5\},\{1,3,4\}\}] - \{\#{:}2, \cup{:}\{1,2,3,4,5\}\}$$

for a set with 2 yet unknown elements (sets representing the matrix columns), whose union cover all 5 integers from 1 to 5 (representing the lines). Now, labelling this single variable yields the solution $S = \{\{1,2,5\},\{1,3,4\}\}$.

Our sets approach with attached functions may thus look like a global constraint. In fact, reasoning with functions of set variables can capture in a very natural manner many classes of problems, without jeopardising efficiency. For the union function this is achieved with inferences triggered by a change in the set variable bounds, such as if set of sets X is known to include one more set s_1 (i.e. in_X grows by s_1), then its union must contain s_1, and if set s_2 is definitely excluded from X (removed from $poss_X$), then the elements of s_2 that were the only support for being present in the union must be removed from the union variable. Similarly, if the union function value is really given by a variable, so should the changes of its bounds affect the set variable it is attached to. The cardinality of the elements of X is also taken into account to possibly lead to early failures.

In any of the two *Cardinal* models, only one variable is needed, in contrast to ILP that requires one variable for each "column". Furthermore, with the union function of *Cardinal* only one constraint is needed to state that the universe must be covered, while ILP and the other *Cardinal* model require one constraint for each "line".

6.7.4 Results

To test the different covering approaches, we turned to a digitally available OR library [Beasley 1990] with set covering benchmarks and we picked file *scpcyc06*, a 240x192 problem matrix corresponding to a CYC problem (number of edges required to hit every 4-cycle in a hyper-cube).

The number n of columns that cover all 240 lines ranges from 0 to 192, being 0 trivially impossible and 192 trivially satisfied. We want to find the optimum (minimum) in between these numbers. For our test we tried to find a solution starting with $n=192$ and successively decrementing n until it was impossible (although faster solutions may be obtained with a Branch-and-Bound minimise predicate). Similarly, starting with $n=0$ we incremented it until obtaining a solution. For a given n, we imposed a time limit of a couple of hours, so that we could be left with a lower and an upper bound for the optimum.

In addition to the three proposed models (ILP plus 2 set-based ones), we included for a fair comparison a *Conjunto* version, since the first simple set-based model could also be directly applied on it. All versions were tested under ECLiPSe producing the bounds reported in Table 6.9.

Table 6.9. Optimum bounds

	ILP	*Conjunto*	*Cardinal* (simple)	*Cardinal* (with union function)
lower	9	8	8	50
upper	76	80	78	78
difference	67	72	70	28

First of all we notice that *Conjunto* and *Cardinal* with the same model, achieve the same lower bound, but different upper bounds (*Conjunto* reaches 80 while *Cardinal* is able to reach 78. By the way, *fd_sets* also only reaches 80.)

ILP presents better results than set solvers; but when using the union function of *Cardinal* (the most expressive version), although the upper bound was still larger than ILP's (78 versus 76), the lower bound was significantly better (50 versus 9) and was easily reached thanks to the inferences on the cardinalities of the set elements.

Cardinal with the union function produced the smallest range. Furthermore, considering all versions, we conclude that the optimum is in the range 50 to 76, with the lower bound coming from *Cardinal* with union, and the upper bound from ILP.

Later, we also compared the different set solvers for two other set covering problems from the same library: *scpe1* (50x500 matrix) and *scpclr10* (511x210). While *Conjunto*, *fd_sets* and *Cardinal* behaved similarly, all three reaching the same bounds, *Cardinal* with the union function improved significantly the solutions, as shown in Table 6.10 with the reached ranges.

Table 6.10. Obtained ranges with or without union function

	Simple sets	With union fn.
scpe1	3..12	5..6
scpclr10	3..35	9..35

Ranges obtained using the union function are much smaller (especially in *scpe1*). Differences in execution times are also noteworthy: for instance, in *scpclr10*, value $n=8$ was proven impossible (thus obtaining 9 as best lower bound) in just one hundredth of a second (Pentium II / 450) while simple *Cardinal* required more than one hour to prove it for $n=2$ (yet it was almost twice as fast as *Conjunto*). For the upper bound, the difference is not so large, a solution for $n=35$ being found in 16 seconds with the union function, and in 106 seconds without it.

Comparisons of results were all performed under the domain of (Constraint) Logic

Programming, an area of interest to this thesis where we wanted to improve techniques for problem solving and which has already shown efficiency advantages in tackling a number of problems. Unfortunately, set-covering is not one of them — in fact, it is a typical problem to be solved with a specialised ILP tool (our ILP model relied on the underlying CLP(*FD*) solver to search for a solution with Boolean variables). For an idea on the difference of approaches and plattforms, we tried to solve the three set-covering problems with the ILP model on CPLEX [1988]. *Scpe1* was then solved obtaining 5 as the minimum, and much better ranges were obtained for the other two problems, namely 54..63 for *scpcyc06* and 24..25 for *scpclr10*, before execution aborted for running out of memory.

While recognising that set covering is an old and very studied problem with very efficient specialised ILP tools to solve it, we reckon that set reasoning is a more natural approach to deal with it, and its application with set functions of *Cardinal* considerably improves other constraint approaches. We thus conjecture that a hybridisation of the two approaches with some interaction between solvers could add declarativity to efficiency and even improve it.

6.7.5 Future Research

In the previous section we have seen that it was very easy to express set covering problems using an attached union function to a set (of sets) variable S. To express a set-partitioning problem instead, we need to ensure that the elements of S are pairwise disjoint. This can also be seen as a set function concerning each pair $<s_1,s_2>$ of elements of S (i.e. disjoint(s_1,s_2)). With such an extension, a set-partitioning problem could also be expressed with a single variable declaration, this time with three attached functions (cardinality, union and pairwise disjointness).

Hence, *Cardinal* can be extended with set functions over all pairs or tuples of a variable, in addition to including other optional functions such as the minimum or maximum and more complex ones, and possibly accept user-defined functions and inferences.

Being able to express constraints over all pairs of a set variable, allows us to easily force a minimum distance between any two elements of a set of integers, for example. Of course, in this case, it is enough to force this distance between consecutive elements. Although a set is just a collection of unsorted elements, for efficiency reasons the system should provide facilities to consider consecutive elements and accept constraints between them (in fact, in practice *Cardinal* codes sets as sorted lists). With such facilities one can also easily force a maximum distance between two consecutive elements, which is extremely useful for scheduling applications. For instance, a timetable could be represented by a set variable where two consecutive courses should be close enough so that no big "holes" are generated. Then other constraints could be included such as the whole duration of the timetable (the difference between the maximum and the minimum function) being among some values, and so on.

Testing new set functions in *Cardinal* (or other set solver) on different types of problems thus seems to be an interesting research direction. Also, on the implementation level, a different and more efficient platform than ECLiPSe Prolog should be considered.

6.8 Conclusions

This chapter presented a new general set constraint solver, *Cardinal*, which efficiently solves CSPs modelled with sets by performing a number of inferences over cardinality of sets. *Cardinal* improves significantly existing solvers based on sets, which allows efficient solving of digital circuits problems as well as general problems over sets. In addition, the extension of *Cardinal* to reason on different set functions other than cardinality (e.g. the union function) showed that other applications could benefit from a simple set modelling with such a powerful solver, thus allying declarativity with efficiency.

The next chapter tackles optimisation problems by developing another logic based on sets, and shows its appropriateness in local search to find improved solutions and speed-up reaching of the optimum.

Chapter 7

Test Pattern Optimisation

We have so far addressed a number of logics that extend Boolean logic with symbolic fault-dependency values to encode simultaneously an increasingly number of circuits (with different sets of faults). A generalising logic was even developed to handle an arbitrary number of diagnoses by directly encoding them into sets.

In this chapter we apply this notion of dependency encoding to handle an optimisation problem for TG (hence, we only need to model two circuits: the normal and faulty circuits). After presenting the problem (section 7.1) and describing two different model approaches for it in sections 7.2 (SAT) and 7.3 (5-valued logic), we discuss their limitations in handling optimality in section 7.4. In section 7.5, we develop a more expressive extended logic for a single circuit, that, rather than keeping track of faults, keeps track of sources of unspecified values and their inversion parities. We subsequently add sets of dependencies on specified values to this logic, so as to model a number of alternative test vectors used in local search (section 7.6). In section 7.7 we compare the different models in terms of solution spaces, and in section 7.8 we combine the different logics into a single tool able to optimise test patterns. Finally, in section 7.9, we present results and compare them with an existing and efficient alternative tool before conclusions are summarised in section 7.10.

7.1 Description

As seen in Chapter 3, test patterns may be partially unspecified, allowing the compaction of test sets by merging compatible tests. In general, test vectors produced by a fault-oriented TG algorithm are partially specified, i.e. some input bits may have an unspecified or "don't care" value denoted by an x. Two *compatible* tests can then be combined into one to reduce the size of the test set generated by an ATG system. For instance, tests $0x1$ and $x11$ could be replaced by the single test 011 by a postprocessing operation referred to as *static compaction*. By contrast, in *dynamic compaction* every partially specified vector is processed immediately after its generation so that it will detect additional new faults [Goel and Rosales 1979-80, Abramovici *et al.* 1986], by assigning Boolean values to unspecified PIs.

Usually, dynamic compaction produces smaller test sets than static compaction with less computational effort. Notice, however, that dynamic compaction cannot be used with a fault-independent TG algorithm.

Finding input test vectors that are less specified increases the probability of having more compatible tests so as to further reduce the length of the test set and, consequently, reduce the cost of future testing. Moreover, if a test set is to be implemented on-chip using Built-In Self-Test (BIST) logic, a larger test set may imply a larger chip area thus increasing the system cost.

For some systems, a single test may also be cheaper to apply if only a small number of inputs have to be specified. Hence, in test generation, less specified tests are preferred. Test pattern optimisation then consists in finding, for a given circuit diagnosis, a test pattern with the maximum number of unspecified inputs.

7.2 SAT Approach

After this problem was addressed in [Hellebrand *et al.* 1995] by using a completely heuristic approach, Flores *et al.* [1998a, 1998b] proposed what was apparently the first *formal* model for minimising the number of specified primary input assignments in test generation for single stuck-at faults (SSFs) in combinational circuits. Their Minimum Test Pattern generator (MTP) tool is based on an integer linear programming (ILP) formulation over a propositional satisfiability (SAT) model for test pattern generation.

In this SAT approach for the fault detection problem, each circuit node v is represented by two variables, v^G and v^F, denoting its logic value in the **good** circuit and in the **faulty** circuit, respectively. In addition, another Boolean variable v^S, referred to as the **sensitisation status** of node x, denotes whether v^G and v^F assume different logic values.

To deal with unspecified inputs, a logic variable x (either the v^G or v^F above) should be ternary (with possible values coming from the set $\{0,1,u\}$). Each such variable is represented with two new variables v^0 and v^1 with the interpretation of Table 7.1. The two variables cannot both have value 1, but they can both have value 0 (for an unspecified value).

Table 7.1. Double Boolean variables

v	(v^0, v^1)
0	(1, 0)
1	(0, 1)
u	(0, 0)

The sensitisation status variable v^S is still kept as a simple 0-1 variable, holding value 1 when v^G and v^F differ and both are specified.

The problem specification is then given by a set of Conjunctive Normal Form (CNF) formulae that must be satisfied to correctly find an assignment to PIs that detects the SSF. CNF formulae express, for the good and faulty circuits, each gate with double Boolean variables for its inputs and output. Other CNF formulae express conditions for the fault activation, propagation and detection, as well as node sensitisation and impossible assignments of values (e.g. $\overline{v^0} \vee \overline{v^1}$, to disallow the (1,1) pair).

Subject to these formulae, the optimisation model for finding a minimum-size test pattern is completed with the ILP goal:

$$minimise \quad \sum_{v \in PIs} (v^0 + v^1)$$

In the worst case, each signal line is associated with five Boolean variables ($v^{G,0}$, $v^{G,1}$, $v^{F,0}$, $v^{F,1}$ and v^S), thus representing a potential search space equivalent to using a 2^5–32 valued logic.

The SAT model as described in [Flores *et al.* 1998b] includes, for each node, the constraint that an unspecified good value implies an unspecified faulty value (i.e. $(v^G=u) \Rightarrow (v^F=u)$). This additional constraint is justified by the fact that the sensitisation status v^S of a node can only assume value 1 when both the values of the node in the good and faulty circuits are specified and assume different logic values. Although such extra constraints may ease the task of finding many solutions, they are nevertheless incorrect in the sense that feasible solutions are discarded, as can be shown in a simple example. In the circuit of Figure 7.1, where values for the normal and faulty circuits are shown in the form v_n/v_p as a 9-valued logic, the test pattern $t = 1x$ detects fault $f = a$ s-a-0 but forces a node (the *and*-gate's output) to have an unspecified value in the good circuit and a specified value (0) in the faulty circuit. These values ($u/0$ or $u/1$) are impossible in the SAT model, which converts them in u/u, therefore only accepting completely specified tests (10 and 11) for this circuit.

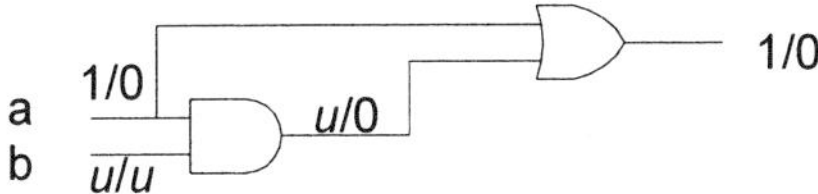

Figure 7.1. Counter-example to a SAT model assumption

So, in practice, this SAT approach can only model 7 values, which constitute a subset of the 9-valued logic as shown in Table 7.2.

Table 7.2. Comparison with 9-valued logic

	$0/0$	$1/1$	$0/1$	$1/0$	u/u	$0/u$	$1/u$	$u/0$	$u/1$
$v^{G,0}$	1	0	1	0	0	1	0		
$v^{G,1}$	0	1	0	1	0	0	1		
$v^{F,0}$	1	0	0	1	0	0	0	(u/u forced)	
$v^{F,1}$	0	1	1	0	0	0	0		
v^{S}	0	0	1	1	0	0	0		

7.3 5-valued Logic

In this section we examine the use of the 5-valued logic (Figure 3.6) already discussed in Chapter 3, to which we add the logic table for S-buffers in Table 7.3.

Table 7.3. S-buffers logic table

Input	S-buffer/0 output	S-buffer/1 output
0	0	$\overline{D}$
1	D	1
D	D	1
$\overline{D}$	0	$\overline{D}$
x	x	x

The 5 values $\{0,1,D,\overline{D},x\}$ provide us with the basic information for each signal line regarding the goal of finding tests for a fault with the maximum number of unspecified inputs. Values 0 and 1 represent constant Boolean values (in the normal and faulty circuits); values D and $\overline{D}$ represent different specified values in the two circuits (1/0 and 0/1 respectively), thus modelling sensitisation; value x represents an unspecified value in either circuit, which allows us to count and maximise the number of unspecified PIs. The optimisation goal is then:

$$maximise \quad \#\{i \in \text{PIs}: i = x\}$$

Table 7.4 shows the relation between the 5 values and the SAT encoding discussed above. The SAT model, while using 5 Boolean variables for each signal line, includes many impossible combinations that must be avoided by explicit formulae. The 4 specified values $\{0,1,D,\overline{D}\}$ of the 5-valued logic correspond to the 4 listed SAT combinations, while the unspecified value x includes all other possible combinations, i.e. those where $v^{G}=(0,0)$ or $v^{F}=(0,0)$, in which case $v^{S}=0$. There are thus only 3 such combinations, namely 00000, 01000 and 10000, since combinations 00010, and 00100 correspond to an unspecified good value and a specified faulty value, which in the SAT model are converted into 00000, i.e. two unspecified values. The 5-valued logic makes no distinctions between partially or completely unspecified composite values v/v_f, (0/u, 1/u, u/0, u/1 or u/u), treating them all as simply unspecified (x).

Table 7.4. Comparison with SAT encoding

	$v^{G,0}$	$v^{G,1}$	$v^{F,0}$	$v^{F,1}$	v^S
0	1	0	1	0	0
1	0	1	0	1	0
D	0	1	1	0	1
$\overline{D}$	1	0	0	1	1
x		other combinations			0

It is then possible to model the test pattern optimisation problem using less information with the more compact 5-valued logic.

7.4 Completeness

When trying to minimise or maximise some function, the goal is, of course, to reach the optimum (usually within a specified limit, such as the maximum allowed computation time). If, however, a model is not complete it may fail to produce the best solution. It is thus important to study the search space corresponding to a specific goal (such as generating, for a target fault, a test with the minimum number of specified PIs) in the adopted model. The search space must not include false solutions and should ideally include the best possible solution. We thus want to know whether a model always allows the optimum to be reached (given enough time to explore the entire search space), i.e. whether the search space always includes the best possible solution. The exhaustion of the search space implies that the best solution (if one was found) is optimal only from the viewpoint of the model. In a complete model, it is definitely the best possible solution; otherwise, we may not ensure it.

An (assumed) limitation of the described SAT model concerns test patterns that do not uniquely identify a set of sensitised paths as in the circuit of Figure 7.2 for the target fault b s-a-1. Assigning 0 to b, PI a may be left unspecified since both $a=0$ and $a=1$ allow the fault effect to propagate to the circuit output (see Figure 7.3). However, the sensitised path (shown in bold) depends on the value of PI a, since by itself the assignment of PI b is not enough to yield any sensitised path.

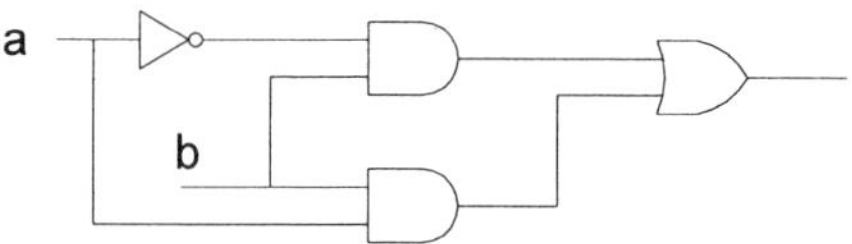

Figure 7.2. Test *x*0 detects *b* s-a-1

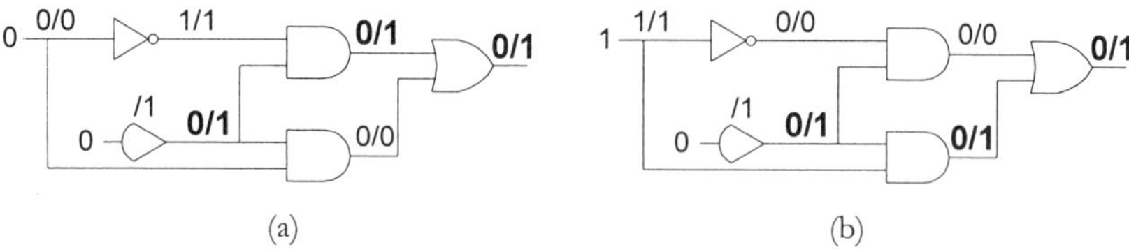

Figure 7.3. Different sensitised paths for different assignments of PI: a) $a=0$; and b) $a=1$

According to [Flores *et al.* 1998b], the SAT model, as well as any other test generation model based on the D-calculus/algorithm [Abramovici *et al.* 1990], is only able to detect tests which guarantee, given the specified PI assignments, propagation of the fault effect to a PO by defining one or more sensitised paths. In fact, the models above, while treating all unspecified values alike, are incomplete and unable to recognise valid tests whose sensitised path depend on the unspecified values (i.e. although the error is assured to be propagated to some PO, its path is not yet certain). Figure 7.4 shows this limitation with the failed attempt of the 5-valued logic to recognise the optimal test of the previous example. Since the output is unspecified (x) we cannot conclude that the test is valid (a PO should hold a *d*-signal D or $\overline{D}$).

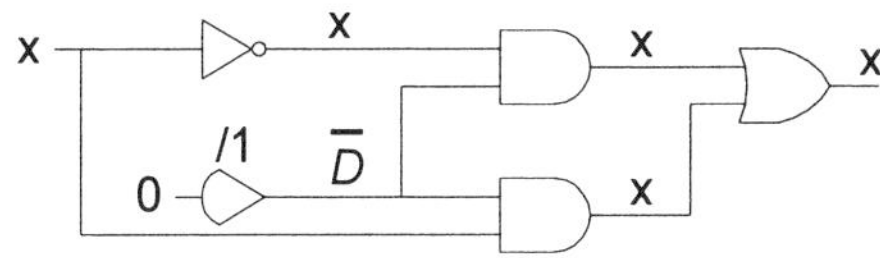

Figure 7.4. 5-valued logic cannot detect test *x*0

Note that this example reproduced in Figure 7.5 for the more complete 9-valued logic, is still not optimally dealt with. The $0/u$ composite value at the output does not allow us to conclude that the test is valid since the value for the faulty circuit remains unspecified (u) instead of holding the required value 1.

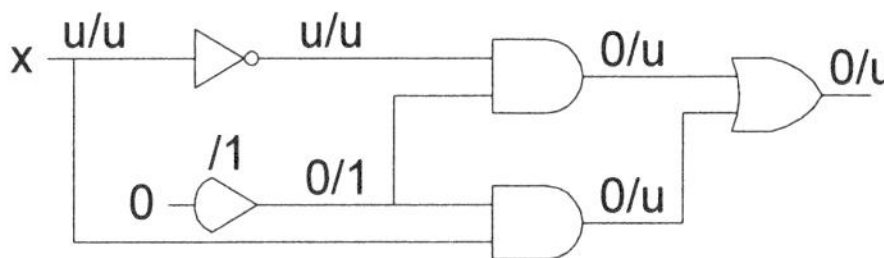

Figure 7.5. 9-valued logic cannot detect test *x*0

7.5 Naming Unspecified Values for an Extended Logic

To correctly handle many such cases, namely for circuits such as the one shown in Figure 7.2, one may extend the logic being used. In this section we propose a method that is able to find the optimal test pattern for such circuits [Azevedo and Barahona 2001].

The difficulty of the example of Figure 7.2 is the reconvergence of an unspecified value. The usual logics, either for a single circuit (e.g. 3-valued logic) or for both the normal and faulty circuits (e.g. 5-valued logic or that implicit in the SAT model), cannot detect that two unspecified values come from the same source (PI a) but have opposite values (i.e. different inversion parities). Therefore, acceptable test patterns may only be found with defined sensitised paths.

To overcome this limitation we attach, to any unspecified value, information regarding its source *id* and its inversion parity *p*. An unspecified value is thus represented as a pair *id-p*. To handle the extra information of an unspecified value, the 3-valued logic (section 2.2) is extended as shown in the *not-*, *xor-* and *and*-tables of Figure 7.6. The negation of an unspecified value has therefore to output a different inversion parity, since it passes through one more inverter. Consequently, when two unspecified values with the same source *id* meet at an *and*-gate with different inversion parities, its output must be zero, since any Boolean assignment to that source will translate into a 0 and a 1 at the *and*-gate's inputs. Similarly, when two such signals meet at an *xor*-gate, the output is necessarily 1. If, on the contrary, they have the same inversion parity then the output is 0 since the two signals have the same value.

NOT	
0	1
1	0
***id*-0**	*id*-1
***id*-1**	*id*-0

AND	0	1	*id*-0	*id*-1
0	0	0	0	0
1	0	1	*id*-0	*id*-1
***id*-0**	0	*id*-0	*id*-0	0
***id*-1**	0	*id*-1	0	*id*-1

XOR	0	1	*id*-0	*id*-1
0	0	1	*id*-0	*id*-1
1	1	0	*id*-1	*id*-0
***id*-0**	*id*-0	*id*-1	0	1
***id*-1**	*id*-1	*id*-0	1	0

Figure 7.6. Extended logic considering the inversion parity of an unspecified value

The application of this logic to the normal and faulty circuits of Figure 7.2 with input vector $v=u0$, is shown in Figure 7.7. In (a), the normal circuit outputs value 0, while in (b) the faulty circuit (with PI b S-buffer s-a-1) outputs value 1 due to the conflicting unspecified values at the *or*-gate. Hence, since it yields two different and specified output values for the two circuits, test vector $v=u0$ indeed detects fault $f = b$ s-a-1.

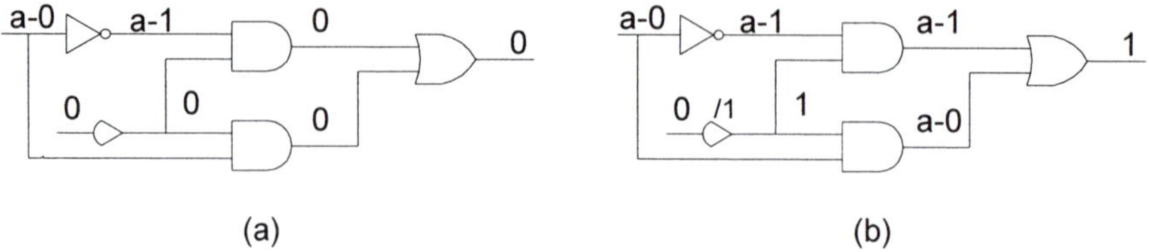

(a)   (b)

Figure 7.7. Normal and faulty circuits with extended logic

This logic extension is sufficient in cases with one unspecified PI (or possibly more, as long as unspecified values with different sources do not meet at some point in the circuit) or, more generally, when there is only one possible unspecified value source. To accept different sources of unspecified values, logic operations between them must be defined since they can meet at some gate. When, say, *and*-ing two such unspecified values $id_A\text{-}p_A$ and $id_B\text{-}p_B$ (with $id_A \neq id_B$), the output Z is also unspecified. Since we want to keep an $id_Z\text{-}p_Z$ structure for the unspecified output Z, and it is not correct to pass any of the inputs to the output as it depends on both inputs, we propose to assign to id_Z the only possible unspecified source: the gate itself. And since it is the source, the inversion parity p_Z must be zero, regardless of p_A or p_B (see Figure 7.8).

$$\text{Id}_A\text{-}p_A \quad\text{—}\quad \boxed{\text{Id}_G} \quad\text{—}\quad \text{Id}_G\text{-}0$$
$$\text{Id}_B\text{-}p_B$$

Figure 7.8. Anding unspecified values of different sources

The *and* operation in this **extended logic** is now complete and is summarised in Table 7.5, remembering that it preserves the ACI (associative, commutative and idempotence) properties.

Table 7.5. Conjunction of named unspecified values

A	B	Z=A.B
0	*Arg*	0
1	*Arg*	**Arg**
Arg	*Arg*	**Arg**
id-0	*id*-1	0
$id_A\text{-}p_A$	$id_B\text{-}p_B$	$id_Z\text{-}0$

Returning to the circuit of Figure 7.2, for which we managed to find the minimum test pattern $t=x0$ with the extended logic, we can make it more complex by turning PI a into a conjunction of two other unspecified values. Still, the fully extended logic easily checks that new test $t_2=xx0$

detects *b* s-a-1 and is minimum since *b* must be assigned with 0. Figure 7.9 represents both the normal and the faulty circuits with the extra *and*-gate. As before, the normal circuit outputs value 0, whereas the faulty circuit outputs the different value 1.

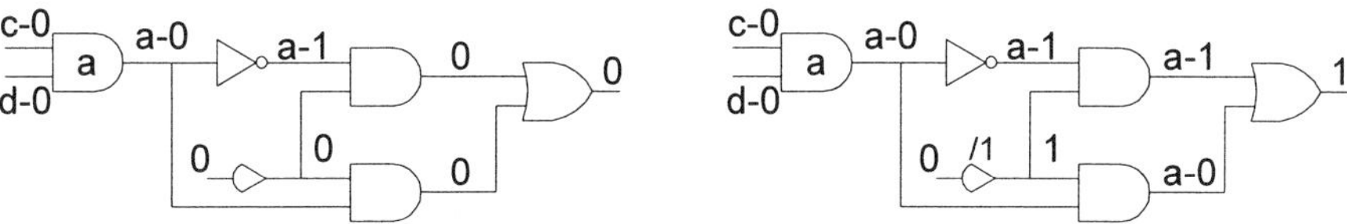

Figure 7.9. Normal and faulty extended circuits

Despite this improvement, if the same logical signal is obtained through different physical gates, the simple dependencies recorded in the values cannot capture this fact, and the model is not able to obtain different output values. This is shown in Figure 7.10, where the initial gate *a* is duplicated.

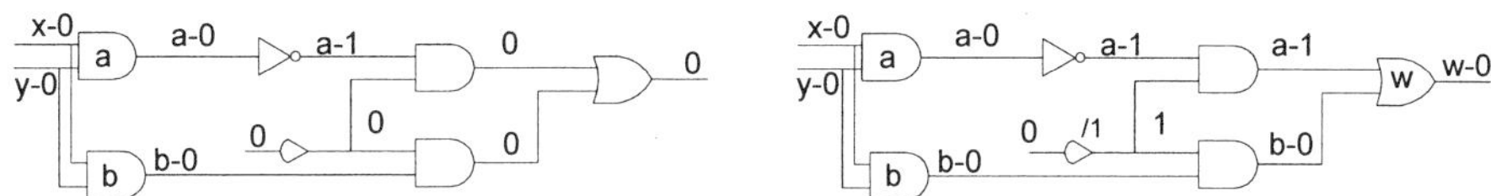

Figure 7.10. Limitation of the extended logic: faulty circuit output should have value 1

Hence, if the conjunction *a* of the unspecified inputs diverges through another gate, as in Figure 7.11 where the PIs take another *and*-gate before the reconvergence at the *or*-gate, then this logic is unable to determine that the final output is 1, leaving it unspecified due to the information loss in logic operations between different unspecified values.

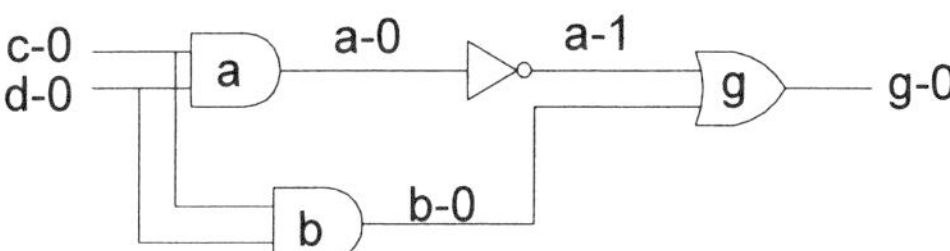

Figure 7.11. Limitation of the extended logic: final output should have value 1

To correctly handle all such cases for an accurate optimality proof, one could have a complete Boolean solver for the unspecified values. Such values should be kept as Boolean formulae with no information loss so as to reach the correct solution as in Figure 7.12. This is, of course, more expensive and brings little practical benefit since similar example cases occur very seldom.

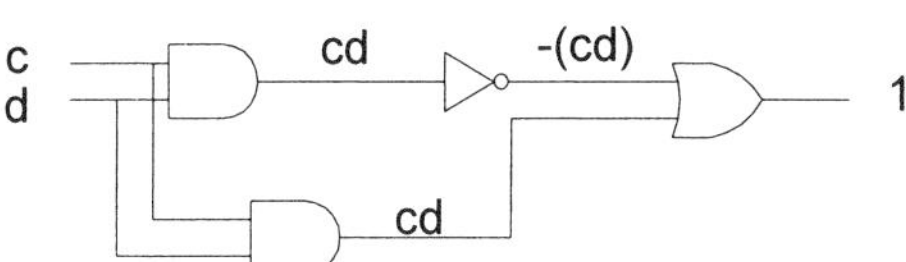

Figure 7.12. Possible Boolean reasoning for unspecified values

Notice that the SAT model described in section 7.2 is insufficient despite its Boolean approach, needed for the unspecified values. The reason is that the SAT model, although using 2 Boolean variables, implicitly implements the 9-valued logic.

7.5.1 Fault Detection Conditions

Denoting by z_n/z_f the values of some bit z, under test t, when the circuit is normal and faulty (with fault f), the example of the previous section only discussed the case where an input pattern forces an output bit z to take either values $z=0/1$ or $z=1/0$. Hence, t detects f if the following proposition holds:

$$\exists z \in POs, \{z_n\} \cup \{z_f\} = \{0,1\}$$

But unspecified normal and faulty values of a PO may still detect the fault, as in Figure 7.13, where if $b=0$ the output is $0/1$ and if $b=1$ the output is $1/0$, being thus always sensitised. Therefore, a test vector still detects a fault if one output bit takes either values $w\text{-}0/w\text{-}1$ or $w\text{-}1/w\text{-}0$, for some source w.

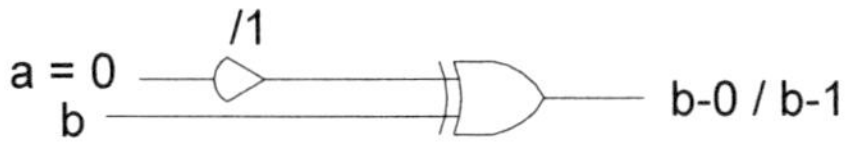

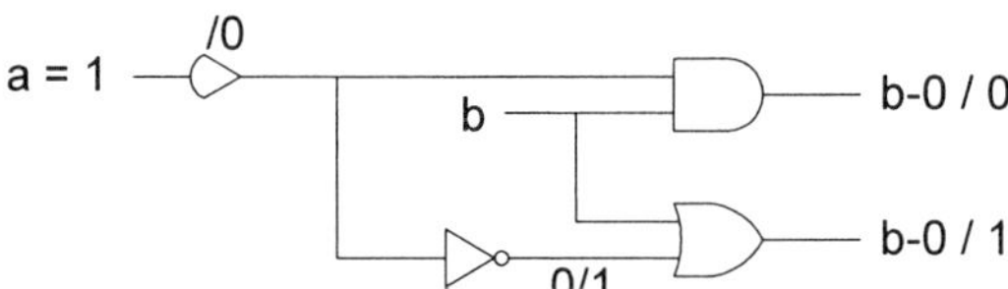

Figure 7.13. Output is sensitised regardless of b

The fault detection condition may then be generalised to:

$$\exists z \in POs, z_n = not_x(zf)$$

where $not_x(v)$ stands for the complement (*not*-gate) operation of value v over the extended logic (see Figure 7.6).

Another, more involved, situation may include more than one sensitised output bit, as in Figure 7.14. Here, one of the two output bits will be sensitised no matter the value of b. If $b=0$, the *or*-gate output will be sensitised to $0/1$; If $b=1$, the *and*-gate output will be sensitised to $1/0$.

Figure 7.14. Different values of b sensitise different POs

Therefore, the extended logic still detects a fault if two output bits take values $w\text{-}0 / 0$ and $w\text{-}1 / 0$ (or any symmetric output, obtained by swapping the normal and faulty values, and/or swapping the 0 and 1). Hence, if from two POs y and z, the 4 values $\{y_n, y_f, z_n, z_f\}$ contain two specified Boolean values $\{B_1, B_2\}$ and two unspecified values with the same source w with parities $\{P_1, P_2\}$, then it is sufficient that the 4 values $\{B_1, B_2, P_1, P_2\}$ have an odd number of 1s. I.e. in general, the fault detection condition for two POs is:

$$\exists y, z \in POs, \{y_n, y_f, z_n, z_f\} = \{B_1, B_2, w - P_1, w - P_2\}, B_1 \oplus B_2 \oplus P_1 \oplus P_2 = 1$$

In addition, if we consider in the circuit two dummy outputs with values $0/0$ and $d\text{-}0/d\text{-}0$, this condition comprises all the above cases.

Exemplifying with a 0/1 PO ($y_n/y_f = 0/1$), with dummy output d-0/d-0 ($z_n/z_f = d$-0/d-0) 2 pairs of values are obtained where $\{y_n, y_f, z_n, z_f\}$ contain 2 Boolean values ($B_1=0$, $B_2=1$) and two unspecified values with the same source d and inversion parity values $P_1=P_2=0$. Hence $B_1 \oplus B_2 \oplus P_1 \oplus P_2 = 1$, i.e. our fault detection condition is set.

Similarly, for a b-0/b-1 PO ($y_n/y_f = b$-0/b-1), dummy output 0/0 ($z_n/z_f = 0/0$) can be used to obtain 2 pairs of values where $\{y_n, y_f, z_n, z_f\}$ contain 2 Boolean values ($B_1=B_2=0$) and two unspecified values with the same source b and inversion parity values $P_1=0$ and $P_2=1$, thus yielding again the fault detection condition $B_1 \oplus B_2 \oplus P_1 \oplus P_2 = 1$.

7.6 Local Search

In the previous section we have studied a simple yet powerful logic to handle unspecified values. The use of a constraint solver over this logic implies large finite domains since any gate is a potential source for an unspecified value. Moreover, two inversion parity values are possible thus doubling the necessary domain size. Although a variable for a circuit PI may have only three values in its domain (the two Boolean values and its own id-0 as possible unspecified source), other variables for the successive gate operations along the circuit will have increasingly larger domains due to the many possible sources of an unspecified value. In addition, since two circuits (normal and faulty) must be simultaneously modelled for TG, domains must be extended to allow all possible combinations and the generalised fault detection condition just discussed would require a complex disjunctive constraint. Such large domains together with the disjunctive goal make constraint reasoning over them very demanding. However, it is straightforward to use this logic to simply **test** alternative solutions, namely those obtained by unspecifying one bit of some already known solution.

An alternative to the constructive approach taken by constraint programming is therefore to use a repairing approach that changes slightly the solution and checks whether this modification still solves the problem. However, in addition to other problems (e.g. traps in local optima), this local search approach relies on a generate-and-test procedure, and may be very inefficient if there is no way to direct the changes into the most promising ones.

7.6.1 A Multiple Extended Logic for Local Search

Despite the positive features of the extended logic to check test patterns, its efficient use on local search depends on an efficient method to find which PIs in a current solution can be unspecified. In this section we develop one such method that takes advantage of this logic for local search over a test pattern in order to improve it by maintaining certain dependencies together with the logic values. Inspection of the output produced by an input test pattern (that is a solution), provides information regarding the bits of such pattern that may change their value to x, thus improving the number of unspecified bits.

The basic idea behind the method is quite simple: if a test t detects fault f, can some specified bit in t become unspecified while still detecting f? To find such bit, one could try to *unspecify* a PI in t and check whether f is still detected. If this simulation does not succeed, one proceeds to the next possible PI, until an unspecifiable PI is found. Unfortunately, such procedure may become cumbersome if there are many possible PIs.

Instead of trying to turn specified bits one at a time, it is possible to make them all possible candidates at start and, with a unique circuit *traversal* (i.e. linearly with the number of circuit gates), check the output to conclude which bits can be indeed unspecified. This is possible by adding dependencies to extended logic values as described below.

Denoting the logic value of signal s after unspecification of PI i by $s \propto i$, let us examine a simple *and*-gate with 2 inputs x and y (Figure 7.15). When two 0's are *and*-ed, the output z is 0 even if one input becomes unspecified (i.e. $z = z \propto x = z \propto y = 0$). For a 01 input combination, the output is still 0 as long as x remains specified (0), i.e. it depends on x. When and-ing two 1's, the output 1 depends on both inputs since if any is unspecified so is the output.

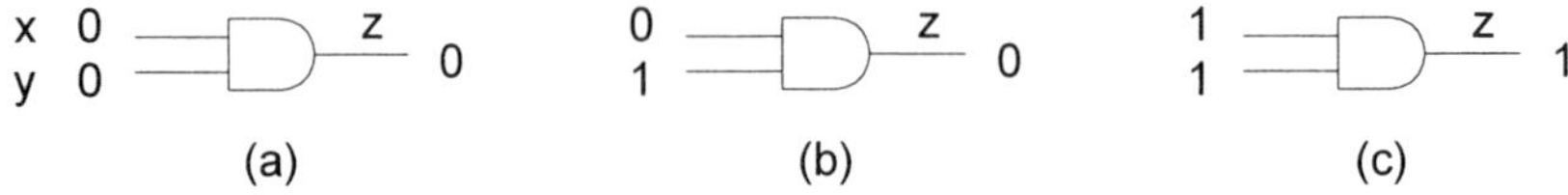

Figure 7.15. Dependencies on *a*) 0, *b*) 1 and *c*) 2 specified values

In Figure 7.15 (b) also note that if x is unspecified, the output assumes the unspecified value of x with its source id and inversion parity 0 ($z \propto x = x\text{-}0$). Whereas for the case of (c) the output is unspecified (as either $x\text{-}0$ or $y\text{-}0$) according to which of the inputs is unspecified. The three dependency cases may thus be represented as in Figure 7.16 where each output signal, in addition to the logic value, has attached a set of conditional unspecified values.

Figure 7.16. Sets of dependencies

A circuit signal s is now represented in the extended form *Set:Value*, where *Value* is, as before, a Boolean value (0/1) or an unspecified value in the form *id-p*. *Set* is a set of conditional values where each member $id_{PI}/Value_{PI}$ means that if PI id_{PI} alone is unspecified, then the physical circuit signal assumes a *different* value $Value_{PI}$ that may also be Boolean or unspecified ($s \propto id_{PI} = Value_{PI}$). Conversely, if no $id_{PI}/Value_{PI}$ is a member of *Set*, then turning solely PI id_{PI} unspecified alone does not affect this circuit signal which assumes value *Value* in both cases ($s = s \propto id_{PI} = Value$).

Actually, since $Value_{PI} \neq Value$, $Value_{PI}$ can never be a Boolean value — it will always be unspecified due to the following

Lemma: When a bit in some input vector t is unspecified, signal lines of the circuit may either be unaffected or also become unspecified ($\forall i \forall s, s \propto i = s \vee \exists id \exists p, s \propto i = id - p$).

The rationale for this is that no extra information is added to the circuit. On the contrary, the circuit becomes less specified, hence no opposite signal values may arise from the fact that a bit has become unspecified, i.e. a signal with value 0 (1) cannot become a signal with value 1 (0).

The proof is easily obtained by contradiction: let us assume that unspecifying PI a (formely 0) flips signal s from 0 to 1. Hence, s would assume value 1 with a unspecified, therefore s is independent of a, i.e. it will assume value 1 whatever the value of a may be. But then, even for the former 0-value of a, s should take value 1. But this is impossible since we had $s=0$ with $a=0$.

The reasoning is identical for different initial Boolean values.

Therefore, if the *Value* of s is unspecified in *Set:Value*, then $Value_{PI}$ will also be so (although possibly with a different source *id*). If *Value* is specified (Boolean) then either $Value_{PI}$ is unspecified or it assumes the same value of *Value*. Of course, in this latter case it will not be in *Set*, since only different values are registered. So, *Set* only contains conditional unspecified values.

7.6.2 Operational Semantics

Let us now use signals in this extended form with dependencies to improve local search. Each signal in the form *Set:Value* represents the extended logic value under the initial test t (given by *Value*) and a set of extended logic values for different input vectors (resulting from the unspecification of some PI in t). Hence, this is a compact representation for circuit signals under t and under all input vectors resulting from the unspecification of any single PI in t.

To compute all these signals, it is necessary to consider PIs with their initial value taken from t. If a PI b is specified with Boolean value *Value* then it is represented by $\{b/b\text{-}0\}$:*Value* since it depends on itself being specified, otherwise *Value* is already unspecified and is then represented by $\{\}$:b-0.

Output signals of the successive gates must be obtained by finding for each gate its output logic value not only under t (as before), but also for all conditional inputs, i.e. for each id_{PI} such that $id_{PI}/Value_{PI}$ is a member of the set of dependencies of at least one of the gate inputs. Each gate then outputs an initial value (the one under t) together with a set of conditional values. This computation must be performed for both the normal and faulty circuits to check the fault detection.

Hence, if in some circuit a gate performs operation op over inputs, say, x and y to output z (i.e. $z = x\ op\ y$), then for each PI i we need to know the value of $z \propto i$.

We first notice that $z \propto i = (x\ op\ y) \propto i = (x \propto i)\ op\ (y \propto i)$, i.e. $\propto$ is compositional with respect to op. Therefore,

- if both x and y are independent of i, so is z and there is no need to explicitly compute $z \propto i$;
- otherwise $(x \propto i)\ op\ (y \propto i)$ is computed to yield $z \propto i$ and if it is different from the original value of z, then $i/\ z \propto i$ is placed in set Set_z of conditional values of z.

Note that value $x \propto i$ $(y \propto i)$ is either present in Set_x (Set_y) in the form $i/\ x \propto i$ $(i/\ y \propto i)$ or is the same as the original x (y).

The *not*-operation is very easy to compute with such logic signals, being the negation (in the extended logic) of the original value and any conditional value in the input set, as shown in Table 7.6.

Table 7.6. Not-operation for local search

NOT	
$S:0$	$\{i/id\text{-}0\colon i/id\text{-}1 \in S\} \cup \{i/id\text{-}1\colon i/id\text{-}0 \in S\} :1$
$S:1$	$\{i/id\text{-}0\colon i/id\text{-}1 \in S\} \cup \{i/id\text{-}1\colon i/id\text{-}0 \in S\} :0$
$S: id_x\text{-}0$	$\{i/id\text{-}0\colon i/id\text{-}1 \in S\} \cup \{i/id\text{-}1\colon i/id\text{-}0 \in S\} : id_x\text{-}1$
$S: id_x\text{-}1$	$\{i/id\text{-}0\colon i/id\text{-}1 \in S\} \cup \{i/id\text{-}1\colon i/id\text{-}0 \in S\} : id_x\text{-}0$

Or, more compactly, the negation of a signal $S{:}V$ is given by $\{i/id\text{-}B\colon i/id\text{-}\overline{B} \in S\} : not_x(V)$, for any Boolean B. Hence, the original value is negated, and the encoded dependencies change parity.

The general rule for computing a gate output in this logic is to find all values for the relevant input vectors (the original one and the ones with a PI turned into unspecified). This procedure is exemplified with the *and*-operation in the pseudo-code of Figure 7.17.

Additionally, we present in Table 7.7, for the *and*-operation, computation rules for the set of conditional values Set_Z of $Z = X\ and\ Y$ as dependent on the original extended logic values of X and Y. Since not all logic values have to be computed and conditional values equal to the original one are not explicitly considered, when translating the general rule to handle signals in the form *Set:Value* the result is a table of different cases for different inputs. In this table, inversion parity values P are Boolean; S_X (S_Y) is the set of conditional values of X (Y); z is the *id* of the *and*-gate; and $S_X \oplus S_Y$ represents the exclusive union of sets S_X and S_Y, which is given by $S_X \oplus S_Y = (S_X \cup S_Y) \setminus (S_X \cap S_Y) = (S_X \setminus S_Y) \cup (S_Y \setminus S_X)$. Wherever unspecified sources id_X and id_Y occur, they are assumed to be different.

```
Procedure And (In: Sx:Vx, Sy:Vy, Out: Sz:Vz);
    Vz ← Vx and Vy        % compute original output value.
    Sz0 ← {}              % start with empty set.
    for each i such that i/_ ∈ Sx∪Sy do   % conditional values.
        i/VXi ∈ Sx or else Vxi ← Vx        % x ∝ i
        i/VYi ∈ Sy or else Vyi ← Vy        % y ∝ i
        VZi ← VXi and VYi                  % z ∝ i
        if VZi ≠ Vz then Sz0 ← Sz0 ∪ {i/VZi}
    end for
    Sz ← Sz0
end Procedure
```

Figure 7.17. *And*-operation procedure for local search

Table 7.7. $Z = X$ *and* Y in local search

original X	original Y	Set$_z$ of ($Z = X$ *and* Y)	original Z
0	0	$\{i/\text{id-P}: i/\text{id-P} \in S_X \cap S_Y\} \cup$ $\{i/\text{z-0}: i/\text{id}_X\text{-P}_X \in S_X \wedge i/\text{id}_Y\text{-P}_Y \in S_Y\}$	0
1	1	$\{i/\text{id-P}: i/\text{id-P} \in S_X \wedge\neg\exists_{\text{idy,Py}} i/\text{id}_Y\text{-P}_Y \in S_Y\} \cup$ $\{i/\text{id-P}: i/\text{id-P} \in S_Y \wedge\neg\exists_{\text{idx,Px}} i/\text{id}_X\text{-P}_X \in S_X\} \cup$ $\{i/\text{z-0}: i/\text{id}_X\text{-P}_X \in S_X \wedge i/\text{id}_Y\text{-P}_Y \in S_Y\}$	1
0	1	$\{i/\text{id-P}: i/\text{id-P} \in S_X \wedge\neg\exists_{\text{idy,Py}} i/\text{id}_Y\text{-P}_Y \in S_Y\} \cup$ $\{i/\text{z-0}: i/\text{id}_X\text{-P}_X \in S_X \wedge i/\text{id}_Y\text{-P}_Y \in S_Y\}$	0
0	id$_Y$-P$_Y$	$\{i/\text{id-P}: i/\text{id-P} \in S_X \cap S_Y\} \cup$ $\{i/\text{id}_Y\text{-P}_Y: i/\text{id}_Y\text{-P}_Y \in S_X \wedge\neg\exists_{\text{id,P}} i/\text{id-P} \in S_Y\} \cup$ $\{i/\text{z-0}: i/\text{id}_X\text{-P}_X \in S_X \setminus S_Y\}$	0
1	id$_Y$-P$_Y$	$\{i/\text{id-P}: i/\text{id-P} \in S_X \cap S_Y\} \cup$ $\{i/\text{id-P}: i/\text{id-P} \in S_Y \wedge\neg\exists_{\text{idx,Px}} i/\text{id}_X\text{-P}_X \in S_X\} \cup$ $\{i/\text{z-0}: i/\text{id}_X\text{-P}_X \in S_X \setminus S_Y\}$	id$_Y$-P$_Y$
id$_X$-P$_X$	id$_X$-P$_X$	$\{i/\text{id-P}: i/\text{id-P} \in S_X \cap S_Y\} \cup$ $\{i/\text{z-0}: i/\text{id-P} \in S_X \oplus S_Y\}$	id$_X$-P$_X$
id$_X$-0	id$_X$-1	$\{i/\text{id-P}: i/\text{id-P} \in S_X \cap S_Y\} \cup$ $\{i/\text{z-0}: i/\text{id-P} \in S_X \oplus S_Y\}$	0
id$_X$-P$_X$	id$_Y$-P$_Y$	$\{i/\text{id-P}: i/\text{id-P} \in S_X \cap S_Y\} \cup$ $\{i/\text{id}_X\text{-P}_X: i/\text{id}_X\text{-P}_X \in S_Y \wedge\neg\exists_{\text{id,P}} i/\text{id-P} \in S_X\} \cup$ $\{i/\text{id}_Y\text{-P}_Y: i/\text{id}_Y\text{-P}_Y \in S_X \wedge \neg\exists_{\text{id,P}} i/\text{id-P} \in S_Y\}$	z-0

As an example, the case in the last line of the table occurs for unspecified inputs of different sources id_X and id_Y, in which case the normal (original) output is, in the extended logic, unspecified z-0. The output signal may assume one of three possible different extended logic values (always unspecified) under the unspecification of PI i:

1. **id-P** (for some id different from id_X, id_Y or z) when i/id-P is a member of both S_X and S_Y (i.e. $x \propto i = y \propto i = id$-P), in which case $z \propto i = (x \propto i)$ *and* $(y \propto i) = id$-P *and* id-P = id-P.

2. **id_X-P_X** (same as original value of X) when i / id_X-P_X is a member of S_Y (i.e. $y \propto i = id_X$-P_X), and there is no i/id-P in S_X (i.e. $x \propto i = id_X$-P_X, as originally), in which case $z \propto i = (x \propto i)$ *and* $(y \propto i) = id_X$-P_X *and* id_X-P_X = id_X-P_X.

3. **id_Y-P_Y** (same as original value of Y) in the symmetric situation (swapping X with Y in the previous case).

The union of all such conditional values yields Set_z, used in a logic signal in the form $Set_z : z\text{-}0$ to improve local search.

7.6.3 Improving Local Search

In Figure 7.18 we return to the example of Figure 7.2 (that other logics could not optimise) with test t=00 that detected b s-a-1 by entailing a 0/1 output. Values for the normal and faulty circuits (b s-a-1) are shown, respectively, above and under each line.

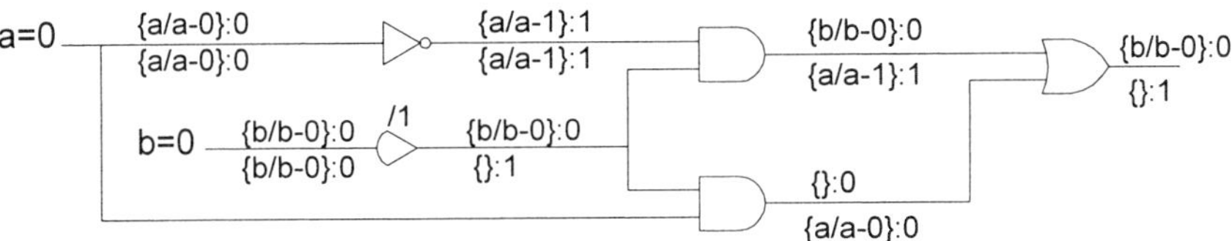

Figure 7.18. Local search logic: PI a may be made unspecified

The faulty circuit output always takes value 1 since if a were unspecified the *or*-gate inputs would be the opposite unspecified values a-1 and a-0. The initial normal 0 output is however dependent on whether b is specified or not. Hence, the fault is detected even if a is unspecified since the detection condition only depends on b (if b is unspecified the output is b-0 / 1). Therefore, the test can be improved to t'=x0 with the output value remaining as 0/1.

In general, a test t for fault f in circuit c can be improved by unspecifying PI a if output vector $Z \propto a$ still satisfies the fault detection condition. By inspecting the circuit POs with their dependency sets after local search, one may conclude that a PI may be made unspecified with no further tests.

Since the original test t already detected fault f and the unspecification of a PI does not bring any new information, we may just look at the originally sensitised POs (i.e. those with a 0/1 or 1/0 original value):

- **If for one such PO z, the conditional value a/V_a is not a member of neither dependency sets (normal or faulty), then $z_n = z_n \propto a$ and $z_f = z_f \propto a$ (or, more compactly, $z = z \propto a$) and PI a may therefore be made unspecified**

- If both z_n and z_f depend on a but remain opposite due to the same source of unspecification with different inversion parities, we have the w-0/w-1 or w-1/w-0 case (i.e. $z_n \propto a = not_x(z_f \propto a)$) and PI a may also be made unspecified.

- If only one of z_n and z_f depends on a, there is still a chance that f may be detected after the unspecification of a, but then we have to look at other POs to check whether a situation such as the one in Figure 7.14 occurs. Then, after unspecification of a, z_n/z_f forms a pair of an unspecified value w-P_1 and a Boolean B_1, and the two values y_n and y_f of another PO y must also include an unspecified value with the same source w-P_2 and a Boolean B_2, to satisfy $B_1 \oplus B_2 \oplus P_1 \oplus P_2 = 1$ as explained in the previous section. These values of $y \propto a$ are easily obtained by direct inspection of the corresponding dependency sets.

This latter situation is exemplified in Figure 7.19 (corresponding to b=0 in Figure 7.14) with a local search over test t=10. Here, PO z was originally sensitised but is dependent on both PIs a and b. However, as indicated in the local search logic values, after unspecification of b we obtain composite values b-0 / 0 and b-0 / 1 for y and z, respectively. Hence, test t'=1x still detects fault f = a s-a-0 due to the general fault detection condition where for POs y, z, we have for the 4 values $\{y_n, y_f, z_n, z_f\}$, 2 Boolean values (B_1=0, B_2=1) and two unspecified values of the same source (b-0, which means $P_1=P_2=0$), yielding $B_1 \oplus B_2 \oplus P_1 \oplus P_2 = 0 \oplus 1 \oplus 0 \oplus 0 = 1$, thus assuring detection of f.

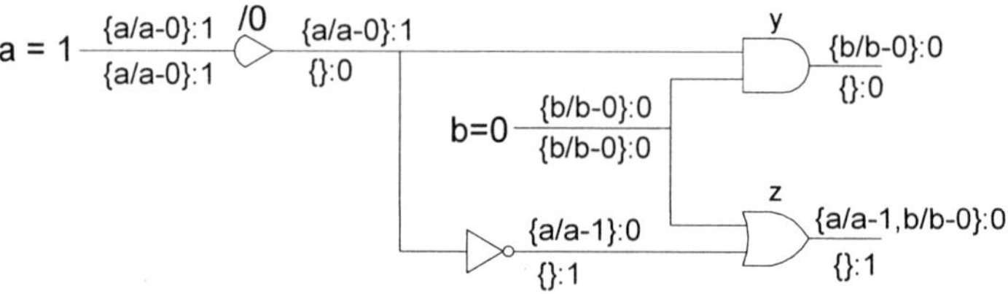

Figure 7.19. PI b can be made unspecified since either y or z will be sensitised

7.6.4 Multiple unspecification

The logic with dependency sets presented above allows one to conclude if a test can be improved by unspecifying one single PI. Let us now examine how it can be used when more than one PI can be made unspecified.

In the example of Figure 7.20, representing ISCAS circuit *c17*, test *t*=00000 detects fault *f*= *23gat s-a-1* (last PO *nand*-gate). Local search over *t* produces the displayed logic values for the normal and faulty circuits. When the normal and faulty values of a signal line are the same, only one value is displayed; otherwise, they are shown above and under the line as in the previous example. Computationally, it is also enough to perform just one logic operation for any gate as long as each of its inputs has two equal values (normal and faulty). Here, only the S-buffer PO presents different values and, by looking at the initial (0/1) and conditional values, we conclude that under this test *t*, only PIs *b* and *e* are crucial and cannot be made unspecified and still detect fault *f*. Hence, any other PI (*a,c,* or *d*) may be made unspecified thus improving the test.

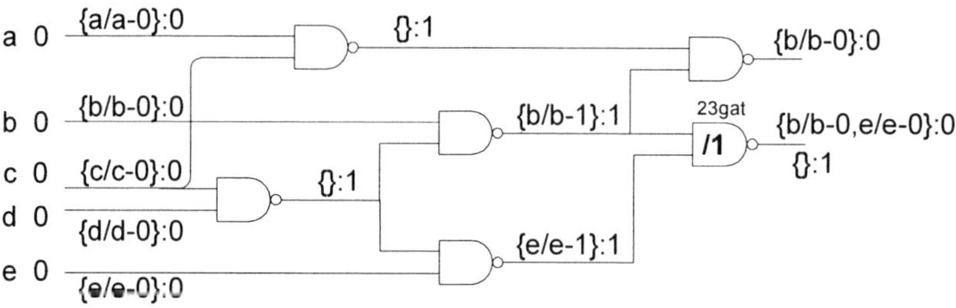

Figure 7.20. PIs b and e cannot be made unspecified in test *t*=00000

It is then assured that unspecifying one of PIs *a*, *c*, or *d* in *t*=00000 produces a still valid test (*x*0000, 00*x*00, or 000*x*0) for fault *f*. Unspecifying more than one of these PIs may also produce valid tests, but this is not guaranteed. For a clearer understanding of the logic for local search, let us check *g*'s output in Figure 7.21 with the extended logic for the previous example where all 3 candidate PIs *a*, *c*, and *d* were simultaneously unspecified. Again, only one logic value is displayed when the values for the normal and faulty circuit are the same. It may seem strange that the value here is the unspecified *g*-1, whereas in Figure 7.20 it was the specified value 1 with no conditional values. The reason for being an "unconditional" value 1 is that it means that unspecifying **one** of the PIs does not change its value. When, as in the case of Figure 7.21, more than one PI is unspecified on the initial test, values cannot be guaranteed (even if independent of any **single** PI of those unspecified). Therefore, such tests with multiple unspecification are not *guaranteed* to still detect the fault.

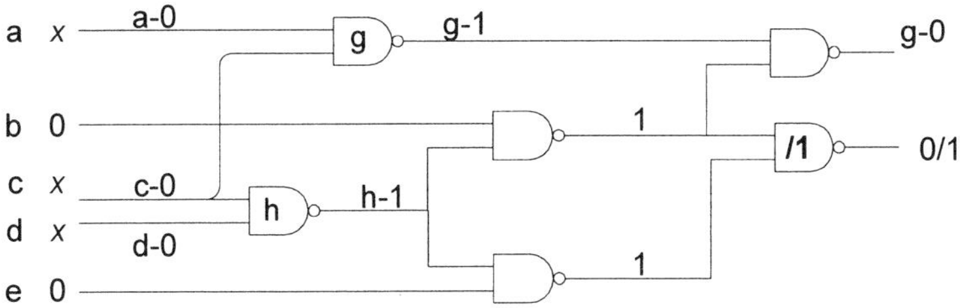

Figure 7.21. Verification of improved test *t'=x0xx0*

One can iteratively improve the test by unspecifying one PI and then perform another local search over the improved test to check whether at least one more PI can be made unspecified, and so forth. However, all candidate PIs can often be simultaneously made unspecified and the optimally improved test remains valid since path sensitisation remains independent of all those input values together, as shown in Figure 7.21. The S-buffer output is the only one to present different values (0/1) and since they are specified at a PO, the fault is still detected. So, when there is a number of candidate PIs, one can use the heuristic of unspecifying all of them and verify whether the resulting test is still valid (this proved to be a good heuristic on experiments, described in section 7.9, where up to around 30 PIs could often be simultaneously made unspecified, thus avoiding a lot of local search).

Hence, although computation of outputs given the inputs of a circuit is not very demanding, significant efficiency improvements can be obtained by directly checking the test with all possible unspecified PIs, since it is not necessary to go through the circuit over and over again. If it is verified that the test does not detect the fault, one can still try with half the possible PIs and then, if it is a valid test, perform another local search, otherwise halve the PIs again until only one is possible.

7.7 Solution Spaces

The models with the logics studied so far are all incomplete when dealing with unspecified values since feasible solutions may be discarded. For completely specified signals, all these models are equivalent, but when there are unspecified inputs, this is not so. It is interesting at this point to compare the logics in terms of the solution spaces of the corresponding models. In the schematic Figure 7.22, *Sol* represents the space of all solutions for a TG problem with possibly unspecified values.

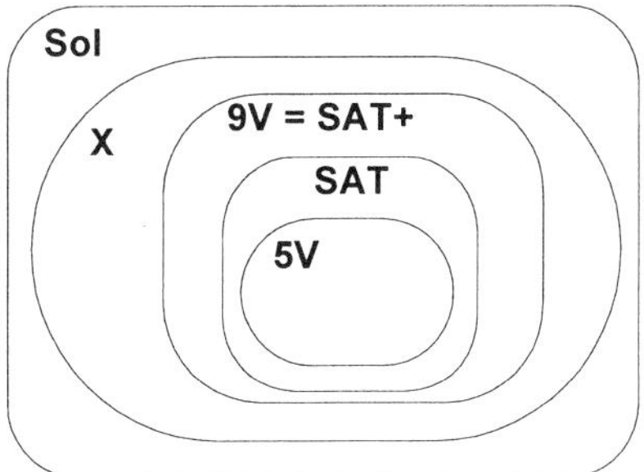

Figure 7.22. Relative solution spaces

In *Sol*, 5 solution spaces relative to other 5 models and their logics are included:

5V   5-valued logic
9V   9-valued logic
X   Extended logic with named unspecified values (section 7.5)
SAT   SAT model shown in section 7.2
SAT+ Complete SAT model, i.e. with the same encoding of SAT but without the limitation that $x^G = u$ implies $x^F = u$

Of course, the sizes of the solution spaces should not be taken rigorously from this diagram. The only relations that it expresses are:

$$Sol(\mathrm{X}) \supseteq Sol(\mathrm{9V}) \supseteq Sol(\mathrm{SAT}) \supseteq Sol(\mathrm{5V})$$
$$Sol(\mathrm{9V}) = Sol(\mathrm{SAT+})$$

where $Sol(Model)$ represents the solution space of *Model*.

These relations have already been addressed in the previous sections. All models treat specified values alike but differ when encoding unspecified values. While we have seen that X is not complete, any unspecified value *id-P* is encoded as *u* in the simpler 9V. Hence, X carries more information. Similarly, any of the three unspecified composite values {*u*/0, *u*/1, *u*/*u*} of 9V is simply encoded in SAT as 00000 (*u*/*u*). Even more "drastically", value *x* of 5V covers all possible unspecified values (i.e. SAT's, 10000, 01000 and 00000, or, more readably, 0/*u*, 1/*u* and *u*/*u*). Therefore, the solution spaces can be regarded as a Russian doll, i.e. one inside the other. Models carrying more information on unspecified values include all solutions of weaker ones and possibly more. This is illustrated with a number of examples that show that a test may be recognised by one of these models and discarded from the other. When applicable, values for the normal and faulty circuits are presented using the v_n/v_f notation. Alternatively, these two values may be shown, respectively, above and under a line as before. In all figures, the weaker model appears on the left with a failed test, whereas a more complete model correctly identifies the test on the right.

In Figure 7.23, test *t*=0*x* for fault *a* s-a-1 is accepted by the SAT model but rejected by 5V. The partially specified composite value 0/*u* at the *and*-gate's output is simply treated as unspecified (*x*) in the 5-valued logic, which given as input to the *or*-gate will also only imply an *x* at the PO, instead of the desired 0/1, correctly computed with the SAT model.

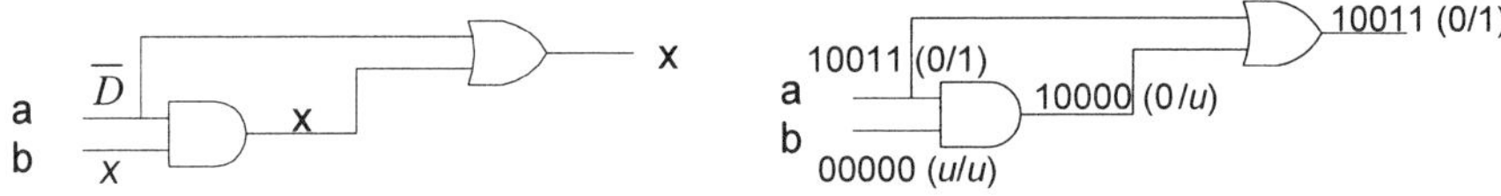

Figure 7.23. Test *t*=0*x* belongs to $Sol(\mathrm{SAT}) \setminus Sol(\mathrm{5V})$

The difference between the two SAT models (SAT and SAT+) is illustrated in Figure 7.24 where test *t*=1*x* for fault *a* s-a-0 is accepted by 9V (=SAT+) but rejected by the incomplete SAT model. In this model the unspecified value (*u*) at the *and*-gate's output in the fault-free circuit forces also an unspecified value in the faulty circuit, which given as input to the *or*-gate will also only imply an unspecified value (*u*) at the PO. Thus, the final partially unspecified composite value 1/*u* (i.e. 01000) does not guarantee sensitisation. This is opposed to 9V that correctly obtains 1/0 at the PO.

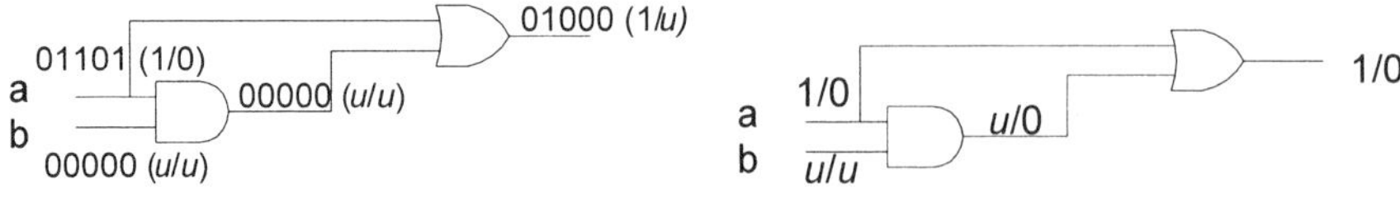

Figure 7.24. Test *t*=*1x* belongs to $Sol(\mathrm{9V}) \setminus Sol(\mathrm{SAT})$

The strength of the extended logic X is shown in Figure 7.25 where it is able to recognise test *t*=*x0* (for *b* s-a-1). The 9V logic (on the left) cannot, due to its impossibility of detecting opposing unspecified values (in this case, *a*-1 and *a*-0 at the faulty circuit's *or*-gate).

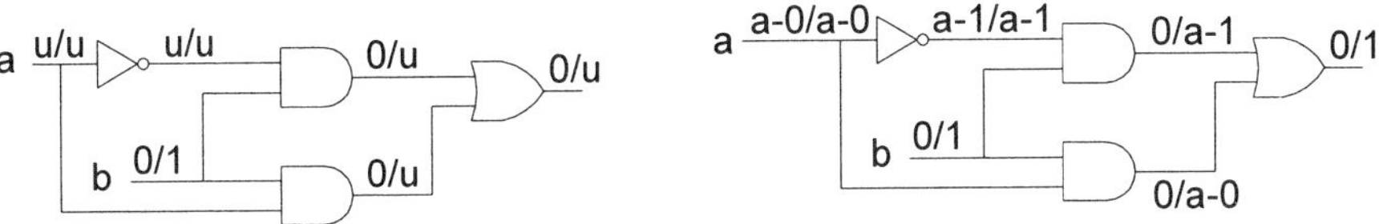

Figure 7.25. Test *t*=*x0* belongs to *Sol*(X)*Sol*(9V)

As seen before, X is still not complete, as illustrated in Figure 7.26 where valid test *t*=*xx0* for *g*/0 is not recognised due the reconvergence of two *and*-ed unspecified values (from *c* and *d*) through 2 and-gates (*a*, *b*), instead of one. The output of normal *or*-gate *g* is given as unspecified (*g*-0), instead of the correct 1-value, because the inputs have two different single sources (*a*, *b*) , instead of the same "double" source (*cd*).

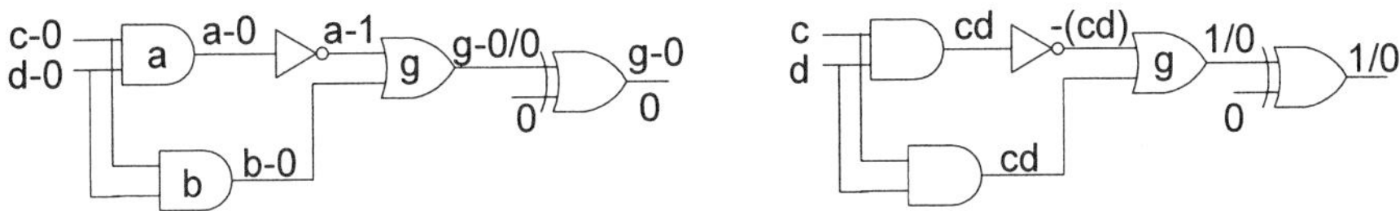

Figure 7.26. Valid test *t*=*xx0* does not belong to *Sol*(X)

7.8 Combining Logics

Optimisation problems may be addressed by constructive or repairing approaches. In this section we show, for this specific problem of test pattern optimisation, an approach that adapts constraint propagation with results obtained from local search in order to outperform the use of each of these techniques alone. The tool that we developed to solve this problem, **Maxx** [Azevedo and Barahona 2001], uses a model based on multi-valued logics and implements a specialised constraint solver, incorporating such adaptation. An interesting characteristic of this adaptation is that constraint propagation and local search do not handle the same modelling of the problem, given the different nature of the constructive and repairing approaches.

A search procedure that produces all the solutions to a problem may be adapted to an optimisation algorithm, by producing all the solutions and choosing the best. This naïve and inefficient method is usually improved by a branch and bound [Papadimitriou and Steiglitz 1982, Balas and Toth 1985] algorithm, available in most Constraint Logic Programming systems. When minimising (maximising) a function, rather than producing all solutions, for every partial solution a minimum (maximum) bound for any solution that completes it is computed. If this bound is worse than the current optimum (an already found solution) the partial solution is abandoned, thus pruning the search space. Of course, branch and bound pruning depends heavily on the ability to compute these bounds.

Unfortunately, such bounds may not be obtained in a simple way in the constraint programming approach to test pattern generation. The function to optimise is the number of unspecified inputs, but the presence of sensitised POs can only be guaranteed when all PIs (or at least those in the transitive fanin of POs under consideration) are instantiated. Although the model of the circuit could be changed in order to make it possible to guarantee that a partially specified input entails a certain output, this improved model is too complex to be of practical use.

For an incomplete solver, checking entailment requires, in general, full enumeration of the input variables, exactly what one wants to avoid !

Alternatively, we use the 5-valued logic that explicitly considers an unspecified input as an extra value of the logic, x, making it possible to count the number of such values in the input pattern. With such logic, bounds on the number of x-values in the solutions obtained by completion of a partial solution may be maintained, enabling the use of a branch and bound algorithm to obtain an optimal solution. Since the logic does not enable the model to cover all possible solutions, it does not guarantee that an optimal solution is found. Nevertheless, such cases are rare and the procedure is quite efficient.

In the previous section we have seen that the extended logic with named unspecified values is particularly suitable for local search by attaching sets of dependencies (on specified PIs) to circuit signals. However, for global search, constraint reasoning over a reduced domain seems to be more appropriate. We thus used constraints over the 5-valued logic since with this domain simplification, constraints can be handled more efficiently while a large majority of valid tests can still be modelled.

Since our local search method requires an already found solution, it is not convenient to address the full test generation problem. However, it might be used to improve the branch and bound, constructive, approach. Whenever a solution (i.e. an input pattern) is obtained, a local search around it (obtained by changing some of the input bits) can be performed to further improve this solution. Such approach hopefully circumvents the inefficiency of changing, by backtracking, the earlier choices made, making it possible to change choices in a non-chronological order.

By integrating these various approaches, we combined these logics to implement a tool, *Maxx*, that aims at maximising the number of unspecified inputs in test sets.

Maxx does so by performing a cycle of local search and branch and bound improvements. For a given test, a local search is performed first to quickly find a good initial bound that is passed to the branch and bound procedure using a constraint solver over the 5-valued logic, so as to maximise the number of PIs with the x value. If the constraint solver finds an improved solution, another local search is performed around it, possibly improving the solution. Control passes back to branch and bound with the new bound. If in a branch and bound phase no improved solution is found within a limited amount of time, *Maxx* exits with the current best solution. If during that time it concludes that no better solution is possible, then *Maxx* exits with an optimal solution in the sense that the 5-valued cannot improve it. Although the proof of optimality is obtained with the 5-valued logic, optimal solutions are not restricted to the solution space of this logic since all solutions pass through a local search using the more expressive extended logic.

Maxx (Figure 7.27) can receive as input a set of diagnoses, each possibly extended with some known test or tagged as redundant. The goal then is to improve the tests, i.e. to find an optimal test for each diagnosis (an SSF or a set of faults).

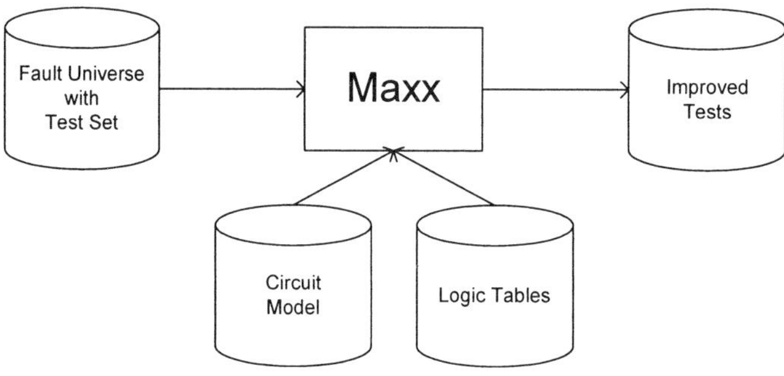

Figure 7.27. *Maxx* system

Given, for some circuit, a test set (possibly empty) referred to as *Base*, *Maxx* tries to improve it as follows: for each possible diagnosis (usually a single fault), if a test is present *Maxx* proceeds as just described above; if the fault is given as redundant then *Maxx* also simply considers it as redundant; otherwise nothing is known about the fault and *Maxx* tries first to find a test for it.

When trying to find an initial test, there is no need to consider unspecified values, since what is important is to know whether some test exists for the fault (or, otherwise, whether it is redundant). After one test is found, one aims at improving it with local search and branch and bound as before. Hence, for faster results, the simpler 4-valued logic of section 3.4.2 may be used together with the Iterative Time-Bounded Search (ITBS) technique of section 3.6. If the search space is completely exhausted with no solution found, the fault is redundant since the model with 4-valued logic is complete for completely specified tests. If no such solution exists then there is also no solution with a partially specified test. If a time limit is reached during the search then it is aborted. If a solution is found we may then proceed to the local search phase to unspecify some PI bits.

Actually, a solution coming from the constraint solver over the 4-valued logic may already contain some unspecified bits in the form of uninstantiated variables. This is due to the labelling strategy that needs only to instantiate those PIs in the transitive fanin of a sensitised PO to ensure that the test is valid. We know that this partially specified input vector entails our desired output since the sensitised PO does not depend on the remaining variables and all possible signals that could affect it are already instantiated, i.e. all the line justification problems for this sensitised path have been solved. This is, nevertheless, a naïve method of checking entailment since this transitive fanin often includes most of, if not all, the PIs. Such method is not sufficient for branch and bound with 4-valued logic, as above discussed, since it does not assure optimal solutions, as many labelled PIs may still be made unspecified and detect the fault (a stronger checking of entailment would be necessary for that).

Figure 7.28 shows a general flowchart of *Maxx*, for each test optimisation, where *LS* stands for Local Search and *BB* stands for Branch and Bound.

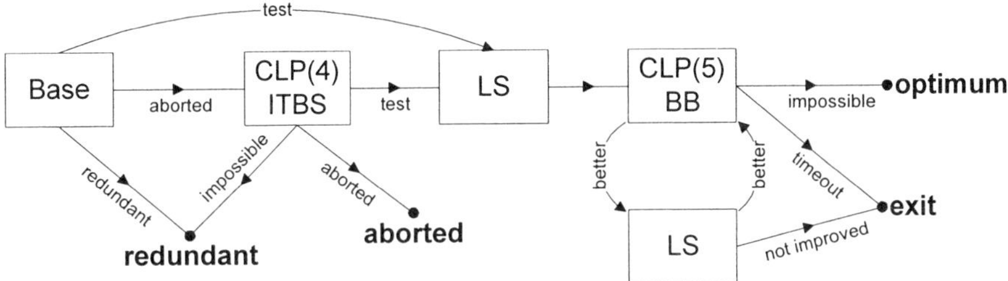

Figure 7.28. Maxx flowchart

In our test, the initial time limit of the ITBS on the solver over the 4-valued logic for aborted faults in the base was 5 seconds. The time limit is consecutively doubled in subsequent rounds until a maximum of 320 ($5*2^6$) seconds (i.e. the maximum continuous search time for each possible *d*-value in the circuit output is restricted to 320 seconds). As to the BB phase with the constraint solver over 5 values, the time allowed to improve a solution was 7 seconds. All these values may nevertheless be parameterised. The propagation strategy used in this 5-valued logic solver was the same as the 4-valued logic solver (described in Chapter 3), with just one more logic value (x) to take into account.

7.9 Results

In this section we present several results of *Maxx* with different base test sets over the ISCAS circuits (summarised again in Table 7.8 — *PI*: primary inputs; *G*: gates; *F*: faults) and compare them with similar tools. All results refer to *Maxx* implemented with ECLiPSe Prolog [ECRC 1994] on a Pentium III, 500 MHz, 256 MB memory. Reported times are in seconds.

Table 7.8. ISCAS circuits: PIs, Gates and Faults

	c17	c432	c499	c880	c1355	c1908	c2670	c3540	c5315	c6288	c7552
PI	5	36	41	60	41	33	233	50	178	32	207
G	6	160	202	383	546	880	1193	1669	2307	2416	3512
F	22	524	758	942	1574	1878	2746	3425	5350	7744	7550

Although *Maxx* was primarily designed to improve a test set, it may also build a test set from scratch if the base test set is empty. Table 7.9 shows the results in this case, where R stands for the number of redundant faults and A, for the aborted ones; X is the total number of unspecified bits in the final test set and $\%X$ its percentage; Opt is the number of optimal solutions from the viewpoint of *Maxx*, as explained before, and $\%Opt$ its percentage; T/f is the average computing time spent per fault.

If F is the number of faults to cover and PI is the number of primary inputs, the length of the test set, *Patterns*, is given by $F - R - A$, and the percentages refer to this number. Then, $\%Opt$ is simply $100*Opt / Patterns$, and $\%X = 100 * X / (Patterns * PI)$.

Table 7.9. *Maxx* results from scratch (empty base test set)

	R	A	X	%X	Opt	%Opt	T/f
c17	0	0	53	48.18	22	100.00	0.02
c432	1	3	13273	70.90	16	3.08	10.42
c499	8	0	8770	28.52	0	0.00	8.45
c880	0	0	47268	83.63	266	28.24	7.17
c1355	8	0	11077	17.25	0	0.00	8.27
c1908	6	2	33201	53.80	24	1.28	9.44
c2670	105	11	577970	94.32	966	36.73	17.85
c3540	128	15	125398	76.42	762	23.22	26.71
c5315	59	0	885245	94.00	1334	25.21	10.32
c6288	34	60	75696	30.92	14	0.18	207.72
c7552	131	2	1362863	88.77	350	4.72	14.77

The toughest circuit, *c6288* (probably due to its much higher level and average fanout, together with a small number of PIs and POs), required a large time per fault due to the aborted faults where, as referred, ITBS is left running at the last round for 320 seconds for each possible output d-value. Moreover, many such output d-values were often possible after setting up circuit constraints, thus multiplying the computation time (up to twice the number of POs, i.e. 2*32=64, since each PO may have both d-0 and d-1 in its domain), which may lead us to think that the ITBS technique may be fine-tuned for some circuits. Still, it is this technique with these parameters that produces the best results, as made clearer in the discussion of the results shown in Chapter 4, devoted to diagnosis.

Since the general idea was to improve a given test set, we picked the results coming from a highly specialised ATG tool, *Atalanta* [Lee and Ha 1993]. *Atalanta*'s goal is to quickly produce a test set with a high fault coverage (see Chapter 3), hence test generation is not fault oriented since a generated test with the aid of powerful heuristics (see section 3.5) may already cover many faults which are immediately discarded. Once a test is found, no improvements on the number of unspecified bits are tried, although there are also heuristics to keep this number high. A user-defined search limit (such as the maximum number of backtrackings) allows *Atalanta* to leave some aborted faults and generate test sets in just a few minutes. Remember that *Maxx* is fault-oriented; uses no special heuristics; is an optimisation (rather than satisfaction) tool and is implemented in Prolog.

Atalanta results are shown in Table 7.10.

Table 7.10. *Atalanta* results

	R	A	X	%X
c17	0	0	48	43.64
c432	3	1	10527	56.23
c499	8	0	5262	17.11
c880	0	0	46471	82.22
c1355	8	0	8556	13.33
c1908	8	0	27589	44.71
c2670	97	20	563624	92.01
c3540	134	0	122854	74.66
c5315	59	0	872143	92.60
c6288	34	387	51976	22.18
c7552	77	181	1311209	86.87

From the tables we see that *Atalanta* already produces some good results in terms of the number of unspecified inputs, although *Maxx* is naturally better for that. We may now see how much can we improve *Atalanta* results by taking it as the base for *Maxx*, in Table 7.11. To ease comparison, *Atalanta* results for the percentage of unspecified bits are reproduced in gray as column $\%X_{Atalanta}$.

Table 7.11. *Maxx* results with *Atalanta* as base

	R	A	X	%X	$\%X_{Atalanta}$	Opt	%Opt	T/f
c17	0	0	53	48.18	43.64	22	100.00	0.01
c432	3	1	13363	71.38	56.23	16	3.08	7.71
c499	8	0	7872	25.60	17.11	0	0.00	7.95
c880	0	0	48154	85.20	82.22	272	28.87	6.08
c1355	8	0	12868	20.04	13.33	0	0.00	7.93
c1908	8	0	32579	52.79	44.71	18	0.96	7.51
c2670	105	11	576664	94.10	92.01	963	36.62	12.48
c3540	134	0	130397	79.24	74.66	755	22.94	7.07
c5315	59	0	881923	93.64	92.60	1104	20.87	7.53
c6288	34	12	69989	28.41	22.18	19	0.25	52.59
c7552	131	0	1396176	90.91	86.87	328	4.42	10.05

Now the number of aborted faults is much smaller and tests considerably better. Also many proofs of redundancy and optimality were obtained. Curiously, in terms of optimality and "quality" of tests, results with an empty base test set often produced better numbers. This means that initial tests generated by *Maxx* with the 4-valued logic are already quite good and compensate the time spent on it. Of course, with a large base, the time per fault decreases especially due to a lower number of aborted faults. Now this time tends to be around 7 seconds which is the branch and bound (BB) time limit of the *Maxx* cycle of local search (LS) plus BB (Figure 7.28), since generally only one cycle is necessary until LS obtains no more improvement. Only two circuits present very different numbers: *c17*, because optimality is almost immediately reached; and *c6288*, because of the remaining aborted faults.

The previously described MTP tool (section 7.2) also tries to improve a given base test set but, in addition, another pre-processing tool, TG-Grasp [Silva and Sakallah 1997], is used for initial aborted faults, so as to obtain a start-up test pattern for MTP. Table 7.12 presents the MTP results over *Atalanta* test sets when allowing 100 conflicts per fault. The values were collected from the original output test sets, which were kindly sent to us by the MTP designers.

Table 7.12. MTP_{100} results over *Atalanta*

	R	A	X	%X	Opt	%Opt	T/f
c17	0	0	52	47.27	22	100.00	n/a
c432	4	0	11376	60.77	0	0.00	3.21
c499	8	0	5759	18.73	0	0.00	4.35
c880	0	0	47361	83.80	110	11.68	2.54
c1355	8	0	8791	13.69	0	0.00	9.12
c1908	8	0	29856	48.38	9	0.48	9.61
c2670	117	0	566103	92.42	598	22.75	10.99
c3540	134	0	127243	77.33	489	14.86	16.81
c5315	59	0	874845	92.89	727	13.74	9.34
c6288	34	0	52874	21.43	55	0.71	36.65
c7552	131	0	1334794	86.92	322	4.34	17.46

A number of remarks have to be made on these results. We first note that *c2670* is reported to have 117 redundant faults, when from Table 7.11 we can conclude that there may be at most 116 such faults. The reason for this is that fault '2007'/0 is considered as redundant in MTP, while *Maxx* is able to find a test for it in a couple of seconds. An example test for it is given below:

```
xxxxxxxx1xx1xxxxxx100100000xxxxxxxxxxxxxxxxxxxxxxxxxxxxxxxxxxxxx
0x00000000xxxxxxxxxxxxxxxxxxxxxxxxxxxxxxxxxxxxxxxxxxxxxxxxxxxxxxx
xxxxxxxxxxxxxxxxxxxxxxxxxxxxxxxxxxxxxxxxxxxxxxxxxxxxxxxxxxxxxxxxx
xxxxxx0xx0xx0x11x111111111x11111xxxxxxxxxxxxxxxx
```

The test was confirmed with different simulators, both of our own and by others commercially available. Interestingly, we may add that *Atalanta* is not able to find a test for it, even after a week of computation.

One could argue that MTP considered the fault to be redundant due to its incompleteness, as discussed before, but MTP is only incomplete when dealing with unspecified values. Finding a completely specified test is sufficient to prove that the fault is not redundant, and MTP is theoretically able to recognise the above test by replacing the unspecified values for Boolean ones.

Hence, the numbers of redundant faults and the absence of aborted faults can not be taken rigorously. Also, the time to find an initial test is not taken into account since it comes either from *Atalanta* or from TG-Grasp.

In the MTP results, the notion of optimality is again relative. If complete, it should correspond to the 9-valued logic. But, as we can see for the *c17* circuit, one less unspecified bit was achieved with MTP, although still considering 100% of optimum tests. This disparity occurs for the fault $0_{gat}/1$, corresponding to PI d s-a-1 in Figure 7.29. Here, the optimal test found by *Maxx*, $t=x110x$, is simply discarded by MTP since it does not allow a composite value such as the output $u/1$ of gate g in the circuit (it converts it to u/u as explained in section 7.2), thus losing the possibility of recognising the sensitised PO. The "optimal" test given by MTP is $t'=0110x$, which, by specifying PI a with Boolean value 0, although not enough to remove the "impossible" value $u/1$ from g's output, allows the first PO to have value $1/0$ instead of $1/u$.

Other strange results occurred with *c6288* where a few tests were not recognised by a 9-valued logic simulator, being only recognised by the extended logic, which the MTP model does not have the power to recognise.

Despite these difficulties, the MTP test sets were tested and its validity verified, so we compared its results with ours in Table 7.13, having both *Atalanta* as base.

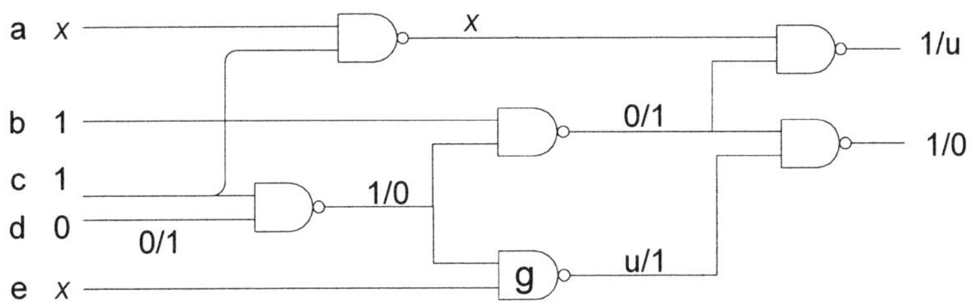

Figure 7.29. *Maxx* optimal test not recognised by MTP

Table 7.13. MTP and *Maxx* improvements on *Atalanta*

	Atalanta	MTP			Maxx			
	%X	%X	Gain	T/f	%X	Gain	T/f	Diff
c17	43.64	47.27	6.44	n/a	48.18	8.06	0.01	1.61
c432	56.23	60.77	10.37	3.21	71.38	34.66	7.16	24.29
c499	17.11	18.73	1.95	4.35	25.60	10.24	7.95	8.29
c880	82.22	83.80	8.89	2.54	85.20	16.76	6.08	7.87
c1355	13.33	13.69	0.42	9.12	20.04	7.70	7.93	7.28
c1908	44.71	48.38	6.64	9.61	52.79	14.63	7.51	7.99
c2670	92.01	92.42	5.13	10.99	94.10	26.16	5.90	21.03
c3540	74.66	77.33	10.54	16.81	79.24	17.92	7.07	7.38
c5315	92.60	92.89	3.92	9.34	93.64	13.51	7.53	9.59
c6288	22.18	21.43	-0.96	36.65	28.41	7.99	12.51	8.96
c7552	86.87	86.92	0.38	17.46	90.91	30.69	10.17	30.31

Assuming that the resources spent in testing the circuit are proportional to the number of *specified* bits, we show the *Gain* (in %) for MTP and *Maxx* wrt *Atalanta* given by $(X_{MTP}-X_{Atalanta})$ / $(100-X_{Atalanta})$, i.e. the saving of resources that can be achieved with the solutions provided by these tools, as well as their difference, *Diff*. The average time spent per fault in each tool, *T/f*, is also shown (for a fairer comparison, the time required to find an initial test was subtracted from *Maxx* times).

Because of different numbers of aborted faults in some circuits between the different tools, which implies different number of test patterns, the percentage of unspecified bits may not be absolutely comparable (as in the negative gain of MTP over its base, for *c6288*). Still, the effect is not significant and the results show that the *Gain* obtained with *Maxx* is always better than those obtained with MTP, with similar times spent per fault in both tools, the algorithms being executed in computers with similar characteristics (MTP in a SUN SPARC, 166 MHz and 384 MB memory, where the slowest SPARC clock is compensated by its RISC architecture and larger memory).

Additionally, we have used MTP as base for *Maxx* to check the improvements. The results that *Maxx* thus obtained are shown in Table 7.14. MTP results, which served as base, are reproduced in gray in column $\%X_{MTP}$ for the percentage of unspecified bits.

The number of redundant faults in *c2670* is kept as 117 since *Maxx* used the MTP results.

Maxx considerably increased the number (and percentage) of unspecified bits in all circuits while keeping the time per fault at an average of 7 seconds (BB time limit). So, from this table we verify that with a simple round of local search and branch bound of *Maxx* (7 seconds), the tests provided by MTP were still significantly improved.

Table 7.14. *Maxx* results with MTP as base

	R	A	X	%X	%X$_{MTP}$	Opt	%Opt	T/f
c17	0	0	53	48.18	47.27	22	100.00	0.01
c432	4	0	13338	71.25	60.77	16	3.08	7.03
c499	8	0	8077	26.27	18.73	0	0.00	7.08
c880	0	0	48672	86.11	83.80	274	29.09	6.00
c1355	8	0	10589	16.49	13.69	0	0.00	7.44
c1908	8	0	35477	57.49	48.38	19	1.02	7.50
c2670	117	0	576847	94.17	92.42	965	36.71	5.57
c3540	134	0	132613	80.59	77.33	773	23.49	6.95
c5315	59	0	883658	93.83	92.89	1115	21.07	7.49
c6288	34	0	66052	26.77	21.43	21	0.27	12.34
c7552	131	0	1397193	90.98	86.92	346	4.66	9.61

Let us now see whether MTP obtains better results when left more time running. Instead of limiting the number of conflicts in MTP to 100, (as in Table 7.12 that we rewrite as MTP$_{100}$), we obtained solutions with MTP with the number of backtracks per fault increased to 1000, MTP$_{1000}$, in Table 7.15. (Results for larger circuits were not available.)

Table 7.15. MTP$_{1000}$ results over *Atalanta*

	R	A	X	%X	Opt	%Opt	T/f
c17	0	0	52	47.27	22	100.00	n/a
c432	4	0	11599	61.96	9	1.73	27.04
c499	8	0	5987	19.47	0	0.00	33.71
c880	0	0	48356	85.56	378	40.13	22.34
c1355	8	0	9753	15.19	0	0.00	64.86
c1908	8	0	31452	50.97	26	1.39	73.44
c2670	117	0	569424	92.96	659	25.07	83.46

With the time per fault considerably increased let us check whether it was worthy. To test the importance of the time limits and starting points imposed to the execution of both MTP and *Maxx*, we tested the tools in a different way, aiming to compare the improvements to the previous MTP solutions that could be found by MTP and *Maxx*. The results are shown in Table 7.16. For *Maxx*, we considered as starting points the solutions obtained in MTP$_{100}$ and kept the 7 seconds time limit in the branch and bound phase. Here, MTP$_{1000}$ times refer to the difference of time per fault of Table 7.15 to the time of MTP$_{100}$, which corresponds to the time needed to improve the MTP$_{100}$ solutions.

Table 7.16. MTP and *Maxx* improvements on MTP$_{100}$

	MTP$_{100}$	MTP$_{1000}$			Maxx			
	%X	%X	Gain	T/f	%X	Gain	T/f	Diff
c17	47.27	47.27	0.00	n/a	48.18	1.73	0.01	1.73
c432	60.77	61.96	3.03	23.83	71.25	26.71	7.03	23.68
c499	18.73	19.47	0.91	29.36	26.27	9.28	7.08	8.37
c880	83.80	85.56	10.86	19.80	86.11	14.26	6.00	3.40
c1355	13.69	15.19	1.74	55.74	16.49	3.24	7.44	1.51
c1908	48.38	50.97	5.02	63.83	57.49	17.65	7.50	12.63
c2670	92.42	92.96	7.12	72.47	94.17	23.09	5.57	15.96
c3540	77.33	n/a	n/a	n/a	80.59	14.38	6.95	n/a
c5315	92.89	n/a	n/a	n/a	93.83	13.22	7.49	n/a
c6288	21.43	n/a	n/a	n/a	26.77	6.80	12.34	n/a
c7552	86.92	n/a	n/a	n/a	90.98	31.04	9.61	n/a

The results are quite meaningful. The improvements achieved on the MTP_{100} results by *Maxx* are much higher than those obtained with MTP_{1000}. Moreover, they are obtained in a fraction of the time that MTP_{1000} requires (for circuits larger than *c2670* we could not even get results with MTP_{1000}).

Finally we analysed what were the sources of our improvements, namely whether the combination of local search and CP was effective. The results are shown in Table 7.17.

Table 7.17. Breakdown of *Maxx* improvements

	MTP	LS_0	CP_1	LS_1	CP_2	LS_2	CP_3	LS_3	CP_4	LS_4
c17	51	2								
c432	11,376	1,714	247	1						
c499	5,759	1,455	863							
c880	47,361	1,064	239	8						
c1355	8,791	867	931							
c1908	29,856	4,318	1,303							
c2670	566,103	9,073	1,637	30	2	2				
c3540	127,243	4,935	397	35	2	1				
c5315	874,845	7,616	1,184	13						
c6288	52,874	10,749	2,126	303						
c7552	1,334,794	56,052	5,630	628	22	29	26	8	2	2

The MTP column shows the number of unspecified bits obtained in MTP_{100}. An important increase in the number of unspecified bits in the test patterns is obtained purely by local search on the initial MTP_{100} solution (LS_0). Nevertheless, a significant improvement is still obtained by executing a constraint programming branch and bound search with the initial bound updated to that obtained in the local search phase (CP_1). Such solution is still often improved by a subsequent local search (LS_1) and in circuit *c7552*, three additional constraint programming and local search steps (CP_2, LS_2 and so on) still improve the previous solutions.

These results show that although the model used in the constraint-programming step (with 5-valued logic) is less complete than that used in the local search step (extended logic with set dependencies), it does provide a different starting point that enables local search to escape from local optima. When a solution is found by constraint propagation, the extended logic used for local search, which is too complex to be the basis of a CP solver, is applicable to the new solutions and usually improves these solutions quite fast.

7.10 Conclusions

This chapter has shown that using both constraint propagation and local search to solve constraint satisfaction and optimisation problems may outperform the use of each of these techniques alone. Although we show this in a particular problem in a specific domain, we believe that integrating these two techniques is an important improvement in constraint programming and should be further exploited.

The problem that we solved is an optimisation problem, but a similar technique can be used in constraint satisfaction problems if we allow "slightly" inconsistent solutions, which can be improved by some local search optimisation, as done in a different domain [Krippahl and Barahona 1999].

The tool that we developed, *Maxx*, does not fully integrate constraint propagation with local search. In fact, once a better solution is found in local search, the constraint propagation phase would start from scratch, although with an updated bound for the branch and bound algorithm. Better integration would be achieved, if the constraint propagation phase could be suspended to allow some local search, and then resumed with the updated bound. Such integration is not available in *Maxx* since it relied on the built-in branch and bound meta-predicate available in most CP tools.

Although the two approaches could be more fully integrated, we believe that our experience is a case for allowing the usual branch and bound procedure of a constructive approach, available from most CP tools, to be parameterised with a user defined local search procedure. In fact, rather than fixing the problem to our specific application, it would be important that such ability would be made available by the CP tools themselves, enabling such a parameterisation of their branch and bound primitives with some local search procedure. The local search step may even use a different and more complex model than that of CP, as exemplified in this chapter, where different logics were used. In fact, the 5-valued logic allowed a more efficient propagation of constraints, and an extended logic in local search allowed reaching improved solutions that the simpler logic could not recognise. Moreover, local search with a more complete model often allowed to easily obtain a substantial leap in current optimal solutions of CP, thus fastening the branch and bound procedure.

We believe the approach followed can be easily adapted to handle other constraint satisfaction and optimisation problems, namely concerning digital circuits.

C h a p t e r   8

Generalisation, Discussion and Conclusion

In this thesis we covered a number of different problems over digital circuits, which we handled with different multi-valued logics. Such logics were formalised and we developed models and constraint solvers for those problems, generalised to MSFs, and compared different approaches.

In this final chapter, we show (section 8.1) that all developed techniques generalise to logical-based systems modelled as a set of propositional rules. For agents that can be modelled as such systems, the discussed problems are very relevant [Schroeder 1998], with applications such as traffic control, integrity checking of databases, alarm-correlation in cellular phone networks, and general diagnosis of the most diverse systems ranging from an automatic mirror furnace to communication protocols. Afterwards, in section 8.2 we discuss the overall work of this thesis and present main conclusions and directions of future research.

8.1 Generalisation

General diagnosis problems have been widely studied for agents, e.g. [Schroeder *et al.* 1997, Schroeder 1998, Horling *et al.* 1999, Fries and Graham 2000], since in single- and multi-agent systems, incorrect assumptions made on interacting processes (due to a somehow changed environment) are likely to produce undesired results (e.g. a walking robot with people appearing in its path). Agents must then adapt to the current environment by first detecting and diagnosing the causes for some unexpected change, and then act accordingly. For instance, in [Horling *et al.* 1999] an Intelligent Home environment is described where major appliances such as the dishwasher, water heater, air conditioner, etc., are each controlled by an individual autonomous agent. A simple agent interaction relates the dishwasher and the water heater, in that the dishwasher uses the hot water produced by the water heater. Normally the dishwasher assumes that sufficient water is available for it to operate, since that is the water heater's task. However, if this assumption fails due to some malfunction or shortage of resources, and the dishwasher lacks of diagnosis capabilities, then it may poorly wash or even going out of order because of insufficient hot water. Being able to diagnose, the dishwasher could first detect that a necessary resource is missing (through sensors or some form of feedback) and review its assumptions, and later refine and validate the diagnosis after other interactions with the water heater for a coordination protocol over the hot water (for example, if the owners take showers in certain times of the day, hot water may be scarce and the dishwasher should take that into account).

Moreover, agents must be able to plan and react in face of unanticipated events in their world since often they act in dynamic and changing environments [Wilkins *et al.* 1995], where *a priori* prediction of all situations is virtually impossible. As stated in [Wilkins *et al.* 1995], "as events render some current activities obsolete, the agents should be able to modify their plans while continuing activities unaffected by those events". The authors thus present a system for persistent agents with such capabilities, which was used for applications such as military operations, real-time tracking and fault diagnosis.

Hence, there is a wide range of agents' applications for the techniques proposed in this thesis for the different diagnostic-related problems. Although not all agents are eligible to directly use such techniques, the general idea can be explored in future research.

We are generally concerned with systems modelled by means of a set of propositional rules where only the systems' input and output may be observed. This means that although, for the different possible faults, we may assume alternative components for the system model, we may not, in general, directly monitor them because they are not accessible. This was the case dealt with in this thesis (so far) for VLSI combinational circuits, where the propositional theory is embodied in the circuit components (gates) but only the input and output bits are accessible for monitoring. Nonetheless, the approach is more general and can be adapted to other situations (e.g. a system whose models can only be tested by observing its behaviour for some inputs).

Three modelling difficulties may arise when trying to generalise the proposed techniques to theories modelled by means of a set of propositional rules, namely:

1. How to model the internal system from its set of propositional rules ?
2. How to model general "faults" ?
3. How to model alternative behaviours of systems (i.e. different sets of rules) ?

These difficulties come from the fact that we have only considered basic 'gates' and stuck-at faults. In the following, we show how to overcome all three points:

(1) Modelling propositional rules in terms of Boolean gates is trivial since such rules can be logically expressed in, for instance, CNF formulae where each *and-*, *or-*, and *not-*operation is modelled by its corresponding logical gate. A digital circuit may thus model a more generic system.

(2) A general fault in a system corresponds to some altered function such as $a \Rightarrow b$ instead of $a \Leftrightarrow b$. Taking this example, we examine the corresponding truth tables in Table 8.1 and consider their CNF formulae as $a \Leftrightarrow b \leftrightarrow (\overline{a}+b).(a+\overline{b})$ and $a \Rightarrow b \leftrightarrow \overline{a}+b$.

Table 8.1. Altered system function

a	b	$a \Leftrightarrow b$	$a \Rightarrow b$
0	0	1	1
0	1	0	1
1	0	0	0
1	1	1	1

The equivalence function can thus be modelled as illustrated in Figure 8.1 (a). When the $(a+\overline{b})$ input to the final *and-*gate is stuck-at-1, as in Figure 8.1 (b), this function is altered to an implication, since only $\overline{a}+b$ remains.

Figure 8.1. (a) Equivalence function; (b) altered (implication) function by an SSF

In Table 8.1, this stuck-at-1 fault corresponds precisely to the 1-value of the implication function when *ab*=01 (when equivalence is false), from which the CNF formula component $(a+\overline{b})$ was taken. Similarly, a stuck-at-0 fault could be considered in an alternative modelling from a formula in disjunctive normal form. In this case, it would be the $\overline{a}.b$ input of the *or-*gate outputting the implication function $(\overline{a}.\overline{b}+\overline{a}.b+a.b)$ that would be stuck-at-0 to produce the equivalence function $(\overline{a}.\overline{b}+a.b)$.

(3) Modelling alternative behaviours of systems was already exemplified in the last point,

since when modelling a fault, the 'faulty' system is also modelled, which is an alternative to the 'normal' sstem.

We now formalise and generalise such modelling. Three generalisations will be performed, namely for:

a) arbitrary altered functions, which may revert the output in more than one input combination (a line in the truth table);
b) functions with an arbitrary number of output bits
c) an arbitrary number of alternative systems

Truth tables for systems whose behaviour is given by a set of propositional rules with n input bits $\{i_1,i_2,...,i_n\}$ have 2^n entries, each with a corresponding output vector. We first turn each entry index $e = (b_1,b_2,...,b_n)$ into a conjunction (*and*-gate) $c(e) = c_1c_2...c_n$ such that (for each $j \in \{1..n\}$):

$$c_j = \begin{cases} i_j, & b_j = 1 \\ \overline{i_j}, & b_j = 0 \end{cases}$$

To each such *and*-gate $c(e)$, and for each observable output bit o, we place an (S-)buffer (section 2.5) $s(e)_o$ in its output (i.e. the *buffer*'s input is the *and*-gate's output). Then, for PO o, we model a disjunction (*or*-gate) of all such buffers $s(e)_o$. The global circuit is the general model for an arbitrary system[*].

Now, to model a system a with output (vector) function Z_a, each S-buffer $s(e)_o$ for which $Z_a(e)_o = 0$ (i.e. the value of o in Z_a for input e is 0) is treated as stuck-at-0. Hence, a simple set of faults (an MSF) defines a system (theory). An arbitrary number of systems may thus be modelled thanks to the general nature of S-buffers.

With our general circuit model we can use the techniques and logics described in this thesis to solve the different problems presented, since a diagnosis for this circuit defines a system and all techniques developed accept MSFs. One may thus generate tests for a system, diagnose a system, differentiate multiple system models, optimise tests, and so forth.

One may argue that, in practice, the circuit model will be too large since it explicitly considers all input combinations. It does not necessarily have to be so since, generally, many such combinations are useless or can be combined to drastically reduce circuit size. Also, some POs may share parts of the circuit, thus avoiding unnecessary duplication. While we provided theoretical results, circuit reduction techniques are, however, out of the scope of this thesis.

8.2 Conclusion and Research Directions

In this thesis, we covered different satisfaction and optimisation problems regarding general systems with a set of controllable inputs and a set of observable outputs. These problems arise from the fact that systems may exhibit an unintended behaviour, and theories for possible malfunctions must then be modelled or generated. To cope with this, we formalised a number of multi-valued logics that express the dependency of Boolean signals on possible faults, based on an existing idea in the literature. Furthermore, we generalised all these problems and logics to MSFs, which is a significant and useful improvement in the area, where many specialised techniques existed just for SSFs, which were then also used to indirectly handle MSFs.

The disjunctive nature of goals in the logic models of different problems led us to develop a new search technique, Iterative Time-Bounded Search (ITBS), to be applied in the enumeration phase

---

[*] Here we refer to stateless logical-based systems (without memory). Systems with memory must be modelled as sequential circuits, which may then be transformed into combinational, as mentioned before.

of the CSP to take advantage of more informed heuristics from a more constrained sub-problem and, at the same time, to not stick to some disjunct choice. The time-bounded nature of this technique ensures that bad initial choices are not 'fatal' as in pure chronological backtracking, since a 'leap' to other choice is allowed after a certain time period. To make the technique complete we made it iterative so that all choices are eventually tried before concluding unsatisfiability. ITBS was described in Chapter 3 and its adequacy exemplified with the constraint solver over the 4-valued logic used in the test generation problem. It was then systematically used for other problems and its efficiency again confirmed in Chapter 4 with another multi-valued logic for differential diagnosis.

The known 4-valued logic (encoding 2 theories, as shown in Chapter 3), consisting of a generalisation of the Boolean logic (Chapter 2), was generalised to an 8-valued logic (encoding 3 theories), which we properly formalised (Chapter 4). This logic was then itself generalised to a logic over sets and Booleans (Chapter 5) to encode an arbitrary number of theories. In addition, we presented a transformation function (in section 5.5.2) for logic signals that allowed modelling problems using just sets.

Such logics allowed: *a)* modelling problems avoiding the duplication of circuits (theories) of SAT approaches, and *b)* using CLP(*FD*) solvers to efficiently solve them.

The logics involving a small number of values and their solvers may, in a future research, be compared with what could be automatically generated from the specification of such a logic [Apt and Monfroy 1999].

For the logic involving sets, we developed *Cardinal*, a general CLP(*Sets*) solver (Chapter 6). We presented some applications that were naturally expressed in constraint programming over sets. We demonstrated it by testing these applications in *Cardinal*, and comparing its efficiency with other approaches. Namely, a set-based model of differential diagnosis of alternative theories was presented and we discussed how such approach could be used to elegantly solve a number of diagnostic related problems, in contrast to specific Boolean SAT approaches. Experimental results with well known circuit benchmarks in this field showed that *Cardinal* clearly outperformed another widely available set solver, due to extra inferences using the set cardinality function.

The ability to use attached functions (other than cardinality) to set domain variables in constraint programming was then exemplified with set covering applications, where even for ILP competitive results were obtained.

We thus showed how set functions such as cardinality and union become a powerful extension. So far, to our knowledge, only the set cardinality function was used in set solvers such as *Conjunto* and in a somewhat passive way. *Cardinal* showed that extra cardinality inferences and other set functions could definitely improve performance and expressiveness. The extension of these set functions was then discussed, promising a wide application area.

Since for many problems, some logic signals were irrelevant or unknown, we presented logics with one or more extra signals to denote unspecified signals. Nonetheless, we pointed out that another advantage of using constraint solvers was that finite domains could simply contain the specified values, while the uninstantiated variable represents an unspecified signal.

However, if unspecified values are part of the problem and we want to reason on them, then unspecified values must be explicitly considered. This is the case with the optimisation problem of maximising the number of unspecified inputs of a test, dealt with in Chapter 7. For that problem we developed and formalised an extended logic that represents unspecified values associated with some source signal and its inversion parity. Models with such logic become more complete, since solutions formerly discarded (with other logics) can now be easily recognised with the extended logic. In addition, we attached to such signals over the extended logic, sets of dependencies on specified values of a given test, and defined the semantics of logic operations over the resulting composite signals. This formalised a logic for local search around a given test, which definitely indicate (by inspection of the output signals) the set of input signals, one of which can safely be turned into unspecified so that the resulting improved test remains valid.

We incorporated such repairing approach in a branch-and-bound procedure to implement *Maxx*, a tool involving a number of different multi-valued logics that generates better tests (and faster) than an existing efficient tool based on SAT and ILP. This has also shown the usefulness of integrating local search in branch-and-bound, thus directing research to parameterising such constructive approaches with a given repairing one that cooperates with it interactively to further improve partial solutions.

All techniques developed apply to logical –based systems modelled as a set of propositional rules, such as digital circuits, for which we have used a widely studied set of general purpose benchmarks, so that meaningful comparisons with other approaches could be performed for different problems. Nevertheless, future research may focus on different specific agents with their particularities and problems. Application of techniques to real such agents (or to specific benchmarks) can be direct, or it can benefit from some specialisation to take advantage of particular features of different instances of the general problems and subjects.

REFERENCES

[Abadir *et al.* 1988] M. S. Abadir, J. Ferguson, and T. E. Kirkland, *Logic Design Verification via Test Generation*, IEEE Transactions on Computer-Aided Design, Vol. 7, Number 1, pages 138-148, January 1988.

[Abramovici *et al.* 1986] M. Abramovici, J. J. Kulikowski, P. R. Menon, and D. T. Miller, *SMART and FAST: Test Generation for VLSI Scan-Design Circuits*, IEEE Design & Test of Computers, Vol. 3, Number 4, pages 43-54, August 1986.

[Abramovici *et al.* 1990] M. Abramovici, M. A. Breuer, and A. D. Friedman, *Digital Systems Testing and Testable Design*, IEEE Press, 1990.

[Agrawal and Agrawal 1972] V. D. Agrawal and P. Agrawal, *An Automatic Test Generation System for ILLIAC IV Logic Boards*, IEEE Transactions on Computers, Vol. C-21, Number 9, pages 1015-1017, September 1972.

[Aiken 1994] Alexander Aiken, *Set Constraints: Results, Applications and Future Directions*, in Proceedings of the 2nd International Workshop on Principles and Practice of Constraint Programming (PPCP'94), Alan Borning (Editor), Springer, pages 326-335, 1994.

[Akers 1976] S. B. Akers, *A Logic System for Fault Test Generation*, IEEE Transactions on Computers, Vol. C-25, Number 6, pages 620-630, June 1976.

[Akers 1978] S. B. Akers, *Binary Decision Diagrams*, IEEE Transactions on Computers, Vol. C-27, Number 6, pages 509-516, June 1978.

[Alferes *et al.* 2001] J. J. Alferes, F. Azevedo, P. Barahona, C. V. Damásio and T. Swift, *Logic Programming Techniques for Solving Circuit Diagnosis*, Technical Report available at URL http://centria.di.fct.unl.pt/~cd/publicacoes/iclp01.ps.gz, 2001.

[Apt and Monfroy 1999] Krzysztof R. Apt and Eric Monfroy, *Automatic Generation of Constraint Propagation Algorithms for Small Finite Domains*, in Proceedings of the 5th International Conference on Principles and Practice of Constraint Programming (CP'99), Joxan Jaffar (Editor), Springer, pages 58-72, 1999.

[Armstrong 1972] D. B. Armstrong, *A Deductive Method of Simulating Faults in Logic Circuits*, IEEE Transactions on Computers, Vol. C-21, Number 5, pages 464-471, May 1972.

[Azevedo and Barahona 1998] Francisco Azevedo and Pedro Barahona, *Generation of Test Patterns for Differential Diagnosis of Digital Circuits* (Extended Abstract), in Proceedings of the 4th International Conference on Principles and Practice of Constraint Programming (CP'98), M. Maher and J.-F. Puget (Editors), Springer, page 462, 1998.

Full version in Proceedings of the 1998 ERCIM/COMPULOG Workshop on Constraints, K. Apt, P. Codognet and E. Monfroy (Editors), 1998.

[Azevedo and Barahona 1999] Francisco Azevedo and Pedro Barahona, *Benchmarks for Differential Diagnosis*, at URL http://ssdi.di.fct.unl.pt/~fa/differential-diagnosis/benchmarks.html, 1999.

[Azevedo and Barahona 2000a] Francisco Azevedo and Pedro Barahona, *A Constraint Programming Approach to Model ATPG Related Problems*, in Proceedings of the IEEE European Test Workshop (ETW'2000), P. Prinetto *et al.* (Editors), pages 315-316, 2000.

[Azevedo and Barahona 2000b] Francisco Azevedo and Pedro Barahona, *Applications of an

Extended Set Constraint Solver, in Proceedings of the 2000 ERCIM/CompulogNet Workshop on Constraints, 2000.

[Azevedo and Barahona 2000c] Francisco Azevedo and Pedro Barahona, *Differentiating Diagnostic Theories through Constraints over an Eight-valued Logic*, in Proceedings of the 14th European Conference on Artificial Intelligence (ECAI'2000), W.Horn (Editor), IOS Press, Amsterdam, pages 73-77, 2000.

[Azevedo and Barahona 2000d] Francisco Azevedo and Pedro Barahona, *Modelling Digital Circuits Problems with Set Constraints*, in Proceedings of the First International Conference on Computational Logic (CL'2000), John Lloyd *et al.* (Editors), Springer, pages 414-428, 2000.

[Azevedo and Barahona 2001] Francisco Azevedo and Pedro Barahona, *Interaction of Constraint Programming and Local Search for Optimisation Problems*, in Proceedings of the 7th International Conference on Principles and Practice of Constraint Programming (CP'01), LNCS 2239, Toby Walsh (Editor), Springer, pages 554-559, 2001.

[Azevedo and Barahona 2002] Francisco Azevedo and Pedro Barahona, *Constraint Programming and Local Search with Multi-valued Logics for Optimisation of Test Patterns*, in IEEE European Test Workshop (ETW'2002), pages 11-12, 2002.

[Balas and Toth 1985] E. Balas and P. Toth, *Branch and Bound Methods*, in *The Traveling Salesman Problem*, John Wiley & Sons, Essex, England, pages 361-401, 1985.

[Beasley 1990] J. E. Beasley, *OR-Library: distributing test problems by electronic mail*, at URL http://mscmga.ms.ic.ac.uk/jeb/orlib, originally described in Journal of the Operational Research Society, Vol. 41, Number 11, pages 1069-1072, 1990.

[Beldiceanu 1990] N. Beldiceanu, *An Example of Introduction of Global Constraints in CHIP: Application to Block Theory Problems*, Technical Report TR-LP-49, ECRC, Munich, Germany, 1990.

[Beldiceanu and Contejean 1994] Nicolas Beldiceanu and Evelyne Contejean, *Introducing Global Constraints in CHIP*, in Mathematical and Computer Modelling, Vol. 12, pages 97-123, 1994.

[Benhamou 1995] F. Benhamou, *Interval Constraint Logic Programming*, in Constraint Programming: Basics and Trends, LNCS 910, A. Podelski (Editor), Springer, March 1995.

[Bessière and Cordier 1994] Christian Bessière and Marie-Odile Cordier, *Arc-Consistency and Arc-Consistency Again*, in Proceedings of the ECAI'94 Workshop on Constraint Processing, 1994.

[Bessière and Régin 1996] Christian Bessière and Jean-Charles Régin, *MAC and Combined Heuristics: Two Reasons to Forsake FC (and CBJ) on Hard Problems*, in Proceedings of the 2nd International Conference on Principles and Practice of Constraint Programming (CP'96), Eugene C. Freuder (Editor), Springer, pages 61-75, 1996.

[Bessière *et al.* 1995] Christian Bessière, Eugene C. Freuder and Jean-Charles Régin, *Using Inference to Reduce Arc Consistency Computation*, in Proceedings of the 14th International Joint Conference on Artificial Intelligence (IJCAI'95), C. S. Mellish (Editor), Vol. 1, Morgan Kaufmann Publishers, pages 592-598, San, Mateo, California, USA, 1995.

[Birkhoff 1967] G. Birkhoff, *Lattice Theory*, in Colloquium Publications, American National Society, Vol. 25, 1967.

[Borning *et al.* 1989] A. Borning, M. Maher, A. Martindale and M. Wilson, *Constraint Hierarchies and Logic Programming*, in Proceedings of the Sixth International Conference on Logic Programming, MIT Press, pages 149-164, 1989.

[Breuer 1971] M. A. Breuer, *A Random and an Algorithmic Technique for Fault Detection Test Generation for Sequential Circuits*, IEEE Transactions on Computers, Vol. C-20, Number 11, pages 1364-1370, November 1971.

[Breuer and Friedman 1976] M. A. Breuer and A. D. Friedman, *Diagnosis & Reliable Design of Digital Systems*, Computer Science Press, Rockville, Maryland, 1976.

[Brglez and Fujiwara 1985] F. Brglez and H. Fujiwara, *A Neutral List of 10 Combinational Benchmark Circuits and a Target Translator in FORTRAN*, in Proceedings of the IEEE Symposium on Circuits and Systems, July 1985.

[Büttner and Simonis 1987] W. Büttner and H. Simonis, *Embedding Boolean Expressions into Logic Programming*, in Journal of Symbolic Computation, Vol. 4, Academic Press, pages 191-205, 1987.

[Carlsson *et al.* 1997] M. Carlsson, G. Ottosson, B. Carlson, *An Open-Ended Finite Domain Constraint Solver*, in Proceedings of Programming Languages: Implementations, Logics, and Programs, 1997.

[Caseau *et al.* 1999] Yves Caseau, François-Xavier Josset and François Laburthe. *CLAIRE: Combining Sets, Search and Rules to Better Express Algorithms*, in Proceedings of the 16th International Conference on Logic Programming (ICLP'99), pages 245-259, 1999.

[Cha *et al.* 1978] C. W. Cha, W. E. Donath, and F. Ozguner, *9-V Algorithm for Test Pattern Generation of Combinational Digital Circuits*, IEEE Transactions on Computers, Vol. C-27, Number 3, pages 193-200, March 1978.

[Chandra and Patel 1989] S. J. Chandra and J. H. Patel, *Experimental Evaluation of Testability Measures for Test Generation*, IEEE Transactions on Computer-Aided Design, Vol. 8, Number 1, pages 93-98, January 1989.

[Charatonik and Podelski 1996] Witold Charatonik and Andreas Podelski. *The Independence Property of a Class of Set Constraints*, in Proceedings of the 2nd International Conference on Principles and Practice of Constraint Programming (CP'96), Eugene C. Freuder (Editor), Springer, pages 76-90, 1996.

[Cheeseman *et al.* 1991] P. Cheeseman, B. Kanefsky and W. Taylor, *Where the Really Hard Problems are*, in Proceedings of the 12th International Joint Conference on Artificial Intelligence (IJCAI-91), Vol. 1, pages 331-337, 1991.

[Codognet and Diaz 1997] P. Codognet and D. Diaz, *A Simple and Efficient Boolean Solver for Constraint Logic Programming*, in Journal of Automated Reasoning, 1997.

[Console and Frederic 1994] *Model Based Diagnosis*, in Annals of Mathematics and Artificial Intelligence, L. Console and G. Frederic (Editors), Vol. 11, Numbers 1-4, 1994.

[CPLEX 1988] CPLEX, in http://www.cplex.com, 1988.

[Damásio *et al.* 1995] C. Damásio, W. Nejdl, L. M. Pereira and M. Schroeder, *Model-Based Diagnosis Preferences and Strategies Representation with Meta-Logic Programming*, in Meta-Logics and Logic Programming, K.Apt and F. Turini, (Editors), MIT Press, pages 269-311, 1995.

[Davis 1984] R. Davis, *Diagnostic Reasoning Based on Structure and Behaviour*, in Artificial Intelligence, Vol. 24, pages 347-410, 1984.

[de Kleer 1986] Johan de Kleer, *"An Assumption-Based TMS"*, *"Extending the ATMS"*, and *"Problem Solving with the ATMS"*, in Artificial Intelligence, Vol. 28, Number 2, pages 127-224, 1986.

[Dechter 1990] Rina Dechter, *Enhancement Schemes for Constraint Processing: Backjumping, Learning, and*

Cutset Decomposition, in Artificial Intelligence, Vol. 41, pages 273-312, 1990.

[Dechter 1992] Rina Dechter, *Constraint Networks*, in Encyclopaedia of Artificial Intelligence (2nd Edition), Stuart C. Shapiro (Editor), Wiley, New York, pages 276-285, 1992.

[Dechter and Pearl 1988] R. Dechter and J. Pearl, *Network-Based Heuristics for Constraint-Satisfaction Problems*, in Artificial Intelligence, Vol. 34, Number 1, pages 1-38, January 1988.

[Devienne *et al.* 1997] P. Devienne, JM. Talbot and S. Tison, *Solving Classes of Set Constraints with Tree Automata*, in Proceedings of the 3rd International Conference on Principles and Practice of Constraint Programming (CP'97), Gert Smolka (Editor), Springer, pages 62-76, 1997.

[Diaz and Codognet 1993] D. Diaz and P. Codognet, *A Minimal Extension of the WAM for clp(FD)*, in Proceedings of the International Conference on Logic Programming, MIT Press, 1993.

[Dincbas *et al.* 1988] M. Dincbas, P. Van Hentenryck, H. Simonis, A. Aggoun, T. Graf and F. Berthier, *The Constraint Logic Programming Language CHIP*, in Proceedings of the International Conference on Fifth Generation Computer Systems, pages 693-702, 1988.

[Dincbas *et al.* 1990] M. Dincbas, H. Simonis, and P. Van Hentenryck, *Solving Large Combinatorial Problems in Logic Programming*, in Journal of Logic Programming, Vol. 8, pages 74-94, 1990.

[Doyle 1979] J. Doyle, *A Truth Maintenance System*, in Artificial Intelligence, Vol. 12, Number 3, pages 231-272, 1979.

[Dressler 1997] O. Dressler, *Diagnostic Information at your Fingertips*, Progress in Artificial Intelligence, Lecture Notes in Computer Science, Vol. 1323, pages 349-360, 1997.

[ECRC 1994] ECRC, *ECLiPSe (a) user manual, (b) extensions of the user manual*, Technical Report, ECRC, 1994.

[Flores *et al.* 1998a] Paulo F. Flores, Horácio C. Neto, Krishnendu Chakrabarty and João P. Marques-Silva, *A Model and Algorithm for Computing Minimum-Size Test Patterns*, in IEEE European Test Workshop (ETW), pages 147-148, May 1998.

[Flores *et al.* 1998b] Paulo F. Flores, Horácio C. Neto and João P. Marques-Silva, *An Exact Solution to the Minimum-Size Test Pattern Problem*, in IEEE/ACM International Workshop on Logic Synthesis (IWLS), pages 452-470, June 1998.

[Forbus 1984] K. D. Forbus, *Qualitative Process Theory*, in Artificial Intelligence, Vol. 24, pages 85-168, 1984.

[Fries and Graham 2000] Terrence P. Fries and James H. Graham, *An Agent-Based Approach to Robust Switching Between Abstraction Levels for Fault Diagnosis*, in Lecture Notes in Artificial Intelligence, Vol. 1864, pages 303 ff., 2000.

[Fröhlich 1998] Peter Fröhlich, *DRUM-II Efficient Model-Based Diagnosis of Technical Systems*, PhD thesis, University of Hannover, 1998.

[Fröhlich and Nejdl 1997] Peter Fröhlich and Wolfgang Nejdl, *A Static Model-Based Engine for Model-Based Reasoning*, in Proceedings of the 15th International Joint Conference on Artificial Intelligence (IJCAI-97), pages 466-473, 1997.

[Frühwirth 1995] T. Frühwirth, *Constraint Handling Rules*, in Constraint Programming: Basics and Trends, LNCS 910, A. Podelski (Editor), Springer, 1995.

[Fujiwara and Shimono 1983] H. Fujiwara and T. Shimono, *On the Acceleration of Test Generation Algorithms*, IEEE Transactions on Computers, Vol. C-32, Number 12, pages 1137-1144,

December 1983.

[Galton 1987] A. P. Galton, *Temporal Logics and their Applications*, Academic Press, London, 1987.

[Garey and Johnson 1979] M. R. Garey and D. S. Johnson, *Computers and Intractability*, W. H. Freeman, San Francisco, CA, 1979.

[Gaschnig 1979] John Gaschnig, *Performance Measurement and Analysis of Certain Search Algorithms*, PhD thesis available as Technical Report CMU-CS-79-124, Carnegie-Mellon University, Pittsburg, PA, 1979.

[Gelfond and Lifshitz 1988] M. Gelfond and V. Lifshitz, *The Stable Model Semantics for Logic Programming*, in Proceedings of the 5th International Conference on Logic Programming (ICLP'88), R. Kowalski and K. Bowen (Editors), MIT Press, pages 1070-1080, 1988.

[Gent and Walsh 1999] I.P. Gent and T. Walsh, *CSPLib: a Benchmark Library for Constraints*, Technical report APES-09-1999, 1999. Available at http://www-users.cs.york.ac.uk/~tw/csplib.
A shorter version appears in the Proceedings of the 5th International Conference on Principles and Practices of Constraint Programming (CP-99).

[Gervet 1997] C. Gervet, *Interval Propagation to Reason about Sets: Definition and Implementation of a Practical Language*, Constraints International Journal, Vol. 1, Number 3, Kluwer Academic Publishers, pages 191-244, March 1997.

[Gervet 1999] C. Gervet, personal communication in 1999.

[Gierz *et al.* 1980] G. Gierz, K. H. Hoffman *et al.*, *A Compendium of Continuous Lattices*, Springer Verlag, 1980.

[Ginsberg 1993] Matthew L. Ginsberg, *Dynamic Backtracking*, in Journal of Artificial Intelligence Research, Vol. 1, pages 25-46, 1993.

[Ginsberg and Harvey 1990] M. L. Ginsberg and W. D. Harvey, *Iterative Broadening*, in Proceeding of the National Conference on Artificial Intelligence (AAAI), pages 216-220, 1990.

[Godoy and Vogelsberg 1971] H. C. Godoy and R. E. Vogelsberg, *Single Pass Error Effect Determination (SPEED)*, IBM Technical Disclosure Bulletin, Vol. 13, pages 3443-3344, April 1971.

[Goel 1981] P. Goel, *An Implicit Enumeration Algorithm to Generate Tests for Combinational Logic Circuits*, IEEE Transactions on Computers, Vol. C-30, Number 3, pages 215-222, March 1981.

[Goel and Rosales 1979] P. Goel and B. C. Rosales, *Test Generation & Dynamic Compaction of Tests*, Digest of Papers 1979 Test Conference, pages 189-192, October 1979.

[Goel and Rosales 1980] P. Goel and B. C. Rosales, *Dynamic Test Compaction with Fault Selection Using Sensitizable Path Tracing*, IBM Technical Disclosure Bulletin, Vol. 23, Number 5, pages 1954-1957, October 1980.

[Golomb and Baumert 1965] S. W. Golomb and L. D. Baumert, *Backtrack Programming*, in Journal of the ACM, Vol. 12, pages 516-524, 1965.

[Goundan 1978] A. Goundan, *Fault Equivalence in Logic Networks*, Ph.D. Thesis, University of Southern California, March 1978.

[Graetzer 1971] G. Graetzer, *LATTICE THEORY: First Concepts and Distributive Lattices*, W. H. Freeman and Company, 1971.

[Gruning *et al.* 1991] T. Gruning, U. Mahlstedt, H. Koopmeiners, *DIATEST: A Fast Diagnostic Test Pattern Generator for Combinational Circuits*, Proceedings of the IEEE International Conference on Computer-Aided Design (ICCAD91), pages 194-197, 1991

[Harary 1969] Frank Harary, *Graph Theory*, Addison-Wesley, 1969.

[Haralick and Elliott 1980] Robert M. Haralick and Gordon L. Elliott, *Increasing Tree Search Efficiency for Constraint Satisfaction Problems*, in Artificial Intelligence, Vol. 14, Number 3, pages 263-313, 1980.

[Hartanto *et al.* 1996] Ismed Hartanto, Vamsi Boppana, W. Kent Fuchs, *Diagnostic Fault Equivalence Identification Using Redundancy Information and Structural Analysis*, in Proceedings of the IEEE International Test Conference (ITC'96), SRC Publication C96216, pages 294-302, October 1996.

[Hartanto *et al.* 1997] I. Hartanto, V. Boppana, W.K. Fuchs, J.H. Patel, *Diagnostic Test Pattern Generation for Sequential Circuits*, in Proceedings of the 15th VLSI Test Symposium (VTS), Monterey, pages 196-202, April 1997.

[Harvey and Ginsberg 1995] W. D. Harvey and M. L. Ginsberg, *Limited Discrepancy Search*, in Proceedings of the 14th International Joint Conference on Artificial Intelligence, 1995.

[Hayes 1977] J. P. Hayes, *Modeling Faults in Digital Logic Circuits*, Rational Fault Analysis, R. Saeks and S. R. Liberty (Editors), Marcel Dekker, New York, pages 78-95, 1977.

[Heintze and Jaffar 1994] Nevin Heintze and Joxan Jaffar. *Set Constraints and Set-Based Analysis*, in Proceedings of the 2nd International Workshop on Principles and Practice of Constraint Programming (PPCP'94), Alan Borning (Editor), Springer, pages 281-298, 1994.

[Hellebrand *et al.* 1995] S. Hellebrand, B. Reeb, S. Tarnick and H.-J. Wunderlich, *Pattern Generation for a Deterministic BIST Scheme*, in Proceedings of the International Conference on Computer-Aided Design, 1995.

[Horling *et al.* 1999] Bryan Horling, Victor Lesser, Régis Vincent, Anna Bazzan and Ping Xuan, *Diagnosis as an Integral Part of Multi-agent Adaptability*, Technical Report 99-03, Department of Computer Science, University of Massachusetts, January 1999.

[Ibarra and Sahni 1975] O. H. Ibarra and S. Sahni, *Polynomially Complete Fault Detection Problems*, in IEEE Transactions on Computers, Vol. C-24, Number 3, pages 242-249, March 1975.

[ISCAS 1985] ISCAS. *Special Session on ATPG*, Proceedings of the IEEE Symposium on Circuits and Systems, pages 663-698, Kyoto, Japan, July 1985.

[Jattar and Lassez 1987] J. Jattar and J.-L. Lassez, *Constraint Logic Programming*, in Proceedings of the 14th ACM Symposium on Principles of Programming Languages, pages 111-119, 1987.

[Jaffar and Maher 1994] J. Jaffar and M. J. Maher. *Constraint Logic Programming: A Survey*, in Journal of Logic Programming, Vol. 19, Number 20, pages 503-581, 1994.

[Kirkland and Mercer 1987] T. Kirkland and M. R. Mercer, *A Topological Search Algorithm for ATPG*, in Proceedings of the 24th Design Automation Conference, pages 502-508, June 1987.

[Krippahl and Barahona 1999] L. Krippahl and P. Barahona, *Applying Constraint Propagation to Protein Structure Determination*, in Proceedings of the 5th International Conference on Principles and Practice of Constraint Programming (CP'99), J. Jaffar (Editor), Lecture Notes in Computer Science, Vol. 1713, Springer-Verlag, pages 289-302, 1999.

[Kuipers 1989] B. Kuipers, *Qualitative Reasoning with Causal Models in Diagnosis of Complex Systems*,

Readings in Qualitative Reasoning about Physical Systems, D. S. Weld & J. deKleer, (Editors), Morgan Kaufmann Publishers, Chapter 10, pages 257-274, 1989.

[Kumar 1992] Vipin Kumar, *Algorithms for Constraint Satisfaction Problems: A Survey*, in AI Magazine, Vol. 13, Number 1, pages 32-34, 1992.

[Larrabee 1992] T. Larrabee, *Test Pattern Generation Using Boolean Satisfiability*, IEEE Transactions on Computer-Aided Design, Vol. 11, Number 1, pages 4-15, January 1992.

[le Provost and Wallace 1993] T. le Provost and M. Wallace, *Generalized Constraint Propagation over the CLP Scheme*, in Journal of Logic Programming, Vol. 16, pages 319-359, 1993.

[Lee 1959] C. Lee, *Representation of Switching Circuits by Binary Decision Diagrams*, Bell System Technical Journal, Vol. 38, Number 6, pages 985-999, July 1959.

[Levendel 1980] Y. H. Levendel, private communication, 1980.

[Lee and Ha 1993] H. K. Lee and D. S. Ha, *On the Generation of Test Patterns for Combinational Circuits*, Technical Report Number 12_93, Department of Electrical Engineering, Virginia Polytechnic Institute and State University, 1993.

[Lindner and Rosa 1980] C. C. Lindner and A. Rosa, *Topics on Steiner Systems*, Annals of Discrete Mathematics, Vol. 7, North Holland, 1980.

[Lloyd 1988] John Lloyd, *Foundations of Logic Programming*, 2nd Edition, Springer-Verlag, 1988.

[Lueneburg 1989] H. Lueneburg, *Tools and Fundamental Constructions of Combinatorial Mathematics*, Wissenschaftverlag, 1989.

[Mackworth 1977] Alan K. Mackworth, *Consistency in Networks of Relations*, Artificial Intelligence, Vol. 8, Number 1, pages 99-118, 1977.

[Mackworth 1992] Alan K. Mackworth, *Constraint Satisfaction*, in Encyclopaedia of Artificial Intelligence (2nd Edition), Stuart C. Shapiro (Editor), Wiley, New York, pages 285-293, 1992.

[Manquinho and Silva 2000] V. Manquinho and J. Marques Silva, *On Using Satisfiability Based Pruning Techniques in Covering Algorithms*, in Proceedings of the ACM/IEEE Design, Automation and Test in Europe Conference, pages 356-363, March 2000.

[Marques-Silva 1995] João Paulo Marques-Silva, *Search Algorithms for Satisfiability Problems in Combinational Switching Circuits*, Ph.D. Dissertation, EECS Department, University of Michigan, May 1995.

[Menezes and Barahona 1996] F. Menezes and P. Barahona, *Defeasible Constraint Solving*, in Over-Constrained Systems. Selected Papers, M. Jampel, E. Freuder and M. Maher (Editors), Lecture Notes in Computer Science, Vol.1106, Springer-Verlag, pages 151-170, 1996.

[Meseguer 1997] P. Meseguer, *Interleaved Depth-first Search*, in Proceedings of the 15th International Joint Conference on Artificial Intelligence, pages 1382-1387, 1997.

[Montanari 1974] U. Montanari, *Networks of Constraints: Fundamental Properties and Applications to Picture Processing*, Information Science, Vol. 7, Number 2, pages 95-132, 1974.

[Muth 1976] P. Muth, *A Nine-Valued Circuit Model for Test Generation*, IEEE Transactions on Computers, Vol. C-25, Number 6, pages 630-636, June 1976.

[Nadel 1989] Bernard A. Nadel, *Constraint Satisfaction Algorithms*, in Computational Intelligence, Vol. 5, pages 188-224, 1989.

[Ousterhout 1994] John K. Ousterhout, *Tcl and the Tk Toolkit*, Addison-Wesley, 1994.

[Pacholski and Podelski 1997] Leszek Pacholski and Andreas Podelski, *Set Constraints: A Pearl in Research on Constraints*, in Proceedings of the 3rd International Conference on Principles and Practice of Constraint Programming (CP'97), Gert Smolka (Editor), Springer, pages 2-5, 1997.

[Papadimitriou and Steiglitz 1982] C. H. Papadimitriou and K. Steiglitz, *Combinatorial Optimization: Algorithms and Complexity*, Prentice-Hall, Englewood Cliffs, NJ, 1982.

[Peng and Reggia 1991] Y. Peng and J. Reggia, *Abductive Inference Models for Diagnostic Problem-Solving*, Springer-Verlag, 1991.

[Pomeranz and Reddy 1998] I. Pomeranz, S.M. Reddy, *A Diagnostic Test Generation Procedure for Synchronous Sequential Circuits based on Test Elimination*, International Test Conference (ITC98). Washington, D.C., USA, pages 1074-1083, 1998.

[Prior 1957] Arthur N. Prior, *Time and Modality*, Clarendon Press, Oxford, 1957.

[Régin 1994] Jean-Charles Régin, *A Filtering Algorithm for Constraints of Difference in CSPs*, in Proceedings of the Twelfth National Conference on Artificial Intelligence (AAAI-94), pages 362-367, 1994.

[Roth 1966] J. P. Roth, *Diagnosis of Automata Failures: A Calculus and a Method*, IBM Journal of Research and Development, Vol. 10, Number 4, pages 278-291, July 1966.

[Roth *et al.* 1967] J. P. Roth, W. G. Bouricius, and P. R. Schneider, *Programmed Algorithms to Compute Tests to Detect and Distinguish Between Failures in Logic Circuits*, IEEE Transactions on Electronic Computers, Vol. EC-16, Number 10, pages 567-579, October 1967.

[Sabin and Freuder 1994] Daniel Sabin and Eugene C. Freuder, *Contradicting Conventional Wisdom in Constraint Satisfaction*, in Proceedings of the 2nd International Workshop on Principles and Practice of Constraint Programming (PPCP'94), Alan Borning (Editor), Springer, pages 10-20, 1994.

[Schroeder 1998] Michael Schroeder, *Autonomous, Model-Based Diagnosis Agents*, Kluwer Academic Publisher, 168 pages, Hardbound, ISBN 0-7923-8142-4, April 1998.

[Schroeder *et al.* 1997] Michael Schroeder, Iara de Almeida Móra and Luís Moniz Pereira, *A Deliberative and Reactive Diagnosis Agent Based on Logic Programming*, in Proceedings of the {ECAI}'96 Workshop on Agent Theories, Architectures, and Languages: Intelligent Agents {III}, LNAI 1193, Jörg P. Müller *et al.* (Editors), Springer, pages 293-308, Berlin, ISBN 3-540-62507-0, August 1997

[Schulz *et al.* 1988] M. H. Schulz, E. Trischler, and T. M. Sarfert, *SOCRATES: A Highly Efficient Automatic Test Pattern Generation System*, IEEE Transactions on Computer-Aided Design, Vol. 7, Number 1, pages 126-137, January 1988.

[Schulz and Auth 1989] M. Schulz and E. Auth, *Improved Deterministic Test Pattern Generation with Applications to Redundancy Identification*, IEEE Transactions on Computer-Aided Design, Vol. 8, Number 7, pages 811-816, July 1989.

[SICStus 1995] Programming Systems Group of the Swedish Institute of Computer Science, *SICStus Prolog User's Manual*, 1995.

[Silva and Sakallah 1997] J. P. M. Silva and K. A. Sakallah, *Robust Search Algorithms for Test Pattern Generation*, in Proceedings of the Fault-Tolerant Computing Symposium, pages 152-161, June 1997.

[Silva *et al.* 1999] L. G. Silva, L. M. Silveira and J. P. Marques-Silva, *Algorithms for Solving Boolean Satisfiability in Combinational Circuits*, in Proceedings of the IEEE/ACM Design and Test in Europe Conference (DATE), pages 526-530, March 1999.

[Simonis 1989] H. Simonis, *Test Generation using the Constraint Logic Programming Language CHIP*, in Proceedings of the Sixth International Conference on Logic Programming, MIT Press, pages 101-112, 1989.

[Simonis 1992] H. Simonis, *Constraint Logic Programming Language as a Digital Circuit Design Tool*, Thesis, 1992.

[Simonis and Dincbas 1987] H. Simonis and M. Dincbas, *Using Logic Programming for Fault Diagnosis in Digital Circuits*, in German Workshop on Artificial Intelligence (GWAI-87), pages 139-148, Geseke, Germany, September 1987.

[Stallman and Sussman 1977] R. M. Stallman and G. J. Sussman, *Forward Reasoning and Dependency-Directed Backtracking in a System for Computer-Aided Circuit Analysis*, in Artificial Intelligence, Vol. 9, pages 135-196, 1977.

[Sterling and Shapiro 1994] Leon Sterling and Ehud Shapiro, *The Art of Prolog: Advanced Programming Techniques*, 2nd Edition, MIT Press, 688 pages, ISBN 0-262-19338-8H, 1994.

[Timoc *et al.* 1983] C. Timoc, M. Buehler, T. Griswold, C. Pina, F. Scott, and L. Hess, *Logical Models of Physical Failures*, in Proceedings of the International Test Conference, pages 546-553, October 1983.

[Tsang 1993] Edward Tsang, *Foundations of Constraint Satisfaction*, Computation in Cognitive Science Series, Academic Press, ISBN 0-12-701610-4, 1993.

[van Benthem 1995] J. van Benthem, *Temporal Logic*, Logic Programming Series, in Handbook of Logic in Artificial Intelligence and Logic Programming, Vol. 4, Clarendon Press, pages 241-350, Oxford, 1995.

[Van Hentenryck 1989] P. Van Hentenryck, *Constraint Satisfaction in Logic Programming*, Logic Programming Series, The MIT Press, 1989.

[Van Hentenryck and Deville 1991] P. Van Hentenryck and Y. Deville, *The Cardinality Operator: a New Logical Connective and its Application to Constraint Logic Programming*, in Proceedings of the Eighth International Conference on Logic Programming, 1991.

[Van Hentenryck and Dincbas 1986] P. Van Hentenryck and M. Dincbas, *Domains in Logic Programming*, in Proceedings of the Fifth National Conference on Artificial Intelligence (AAAI'86), Philadelphia, PA, August 1986.

[Walsh 1997] T. Walsh, *Depth-Bounded Discrepancy Search*, in Proceedings of the 15th International Joint Conference on Artificial Intelligence (IJCAI-97), 1997.

[Wilkins *et al.* 1995] David E. Wilkins, Karen L. Myers, John D. Lowrance and Leonard P. Wesley, *Planning and Reacting in Uncertain and Dynamic Environments*, in Journal of Experimental and Theoretical AI, Vol. 7, Number 1, Taylor & Francis, pages 197-227, 1995.

[XSB 2000] *The XSB Programmer's Manual: version 2.1*, at URL http://xsb.sourceforge.net, Vols. 1 and 2, 2000.

[Zhou 2000] Neng-Fa Zhou, *Programming Constraint Propagation in Reactive Rules*, in Proceedings of the 1st International Workshop on Rule-Based Constraint Reasoning and Programming, 2000.

Appendices

Appendix A

— ISCAS Circuits —

ISCAS circuits

In this appendix we describe in more detail the ISCAS'85 benchmark circuits used throughout this thesis.

The following tables present for each such circuit, its intended function together with general statistics on its lines and gates.

Circuit	PI	PO	gates	level	avg fanin	max fanin	fanout stems	fanout lines	avg fanout	max fanout
c432	36	7	160	17	2.10	9	89	236	1.75	9
c499	41	32	202	11	2.02	5	59	256	1.81	12
c880	60	26	383	24	1.90	4	125	437	1.70	8
c1355	41	32	546	24	1.95	5	259	768	1.87	12
c1908	33	25	880	40	1.70	8	385	995	1.67	16
c2670	233	140	1193	32	1.74	5	454	1244	1.55	11
c3540	50	22	1669	47	1.76	8	579	1821	1.72	16
c5315	178	123	2307	49	1.90	9	806	2830	1.81	15
c6288	32	32	2416	124	1.99	2	1456	3840	1.97	16
c7552	207	108	3512	43	1.75	5	1300	3833	1.68	15

Circuit	Function	buffer	not	and	nand	or	nor	xor	Total
c432	Priority decoder		40	4	79		19	18	160
c499	ECAT		40	56		2		104	202
c880	ALU and control	26	63	117	87	29	61		383
c1355	ECAT	32	40	56	416	2			546
c1908	ECAT	162	277	63	377		1		880
c2670	ALU and control	196	321	333	254	77	12		1193
c3540	ALU and control	223	490	498	298	92	68		1669
c5315	ALU and selector	313	581	718	454	214	27		2307
c6288	16-bit multiplier		32	256			2128		2416
c7552	ALU and control	534	876	776	1028	244	54		3512

Below we present high-level schematics of each circuit, with brief descriptions of their functions. The information is taken directly from http://www.eecs.umich.edu/~jhayes/iscas/ where a more recent publication on such reverse engineering is referred [M. Hansen, H. Yalcin, and J. P. Hayes, *Unveiling the ISCAS-85 Benchmarks: A Case Study in Reverse Engineering*, IEEE Design and Test, vol. 16, no. 3, pp. 72-80, July-Sept. 1999].

In the web page, the following introductory comment is then found:

"The high-level ISCAS-85 benchmarks discussed in this paper are available below, and we invite other researchers to use them. The models, of which we have constructed both structural and behavioral versions, partition the original gate-level netlists into standard RTL blocks and identify the functions of these blocks. Together, the gate-level and high-level models form a set of hierarchical benchmark circuits that have proven to be useful research tools in several areas of digital design, including test generation, timing analysis, and technology mapping. The web documentation for each model consists of annotated circuit schematic diagrams, and executable (simulatable) descriptions written in structural Verilog. The structural models are intended to express the specific high-level structure implicit in the original gate-level designs. In most cases, we also provide behavioral Verilog models, which define high-level blocks in the form of logical equations that can readily be synthesized into gates."

Hence, this web page can be consulted for a more comprehensive analysis of such circuit benchmarks, including their complete gate-level description.

C432
27-channel interrupt controller

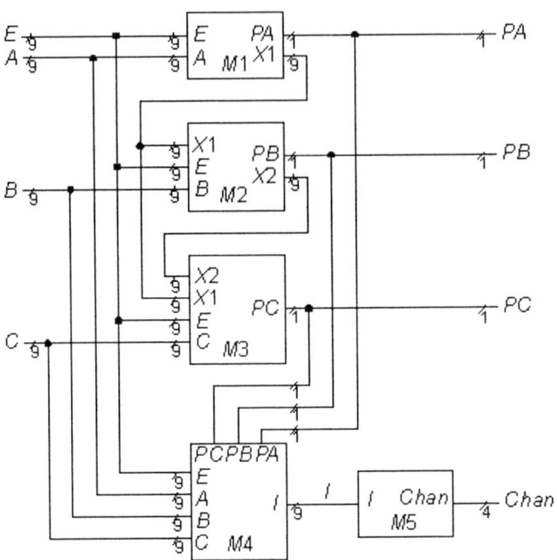

Statistics: 36 inputs; 7 outputs; 160 gates

Function: *c432* is a 27-channel interrupt controller. The input channels are grouped into three 9-bit buses (we call them *A*, *B* and *C*), where the bit position within each bus determines the interrupt request priority. A fourth 9-bit input bus (called *E*) enables and disables interrupt requests within the respective bit positions. The figure above concisely represents the circuit. Modules labelled *M1*, *M2*, *M3*, *M4* contain the underlying logic.

The seven outputs *PA*, *PB*, *PC* and *Chan[3:0]* specify which channels have acknowledged interrupt requests.

Module M1

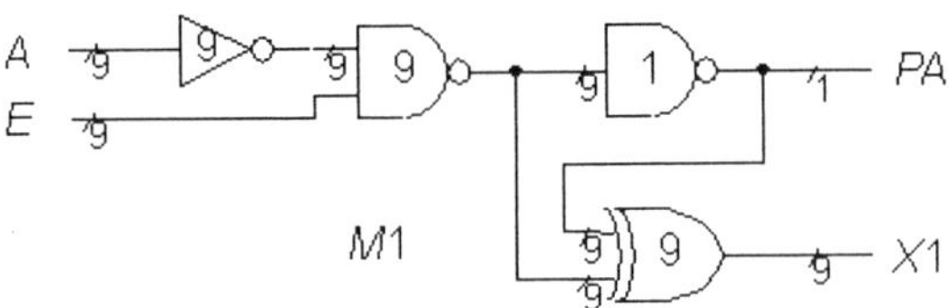

Module M2

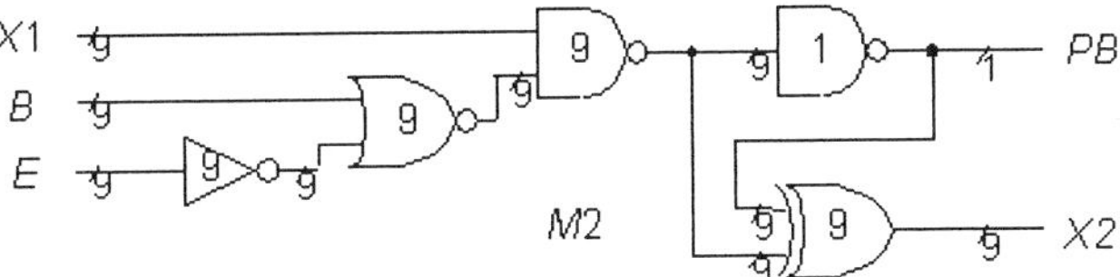

Module M3

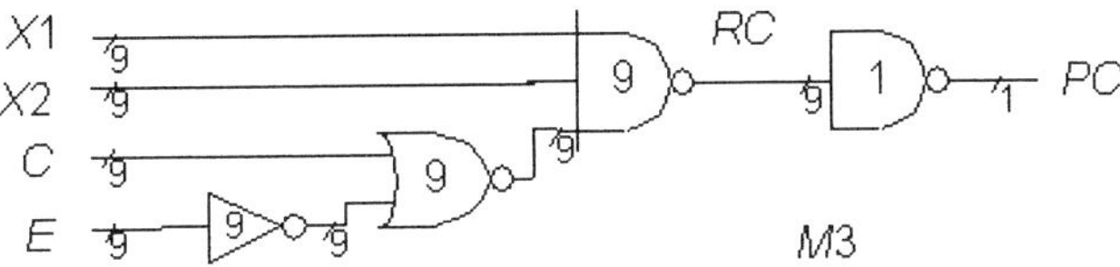

Module M4

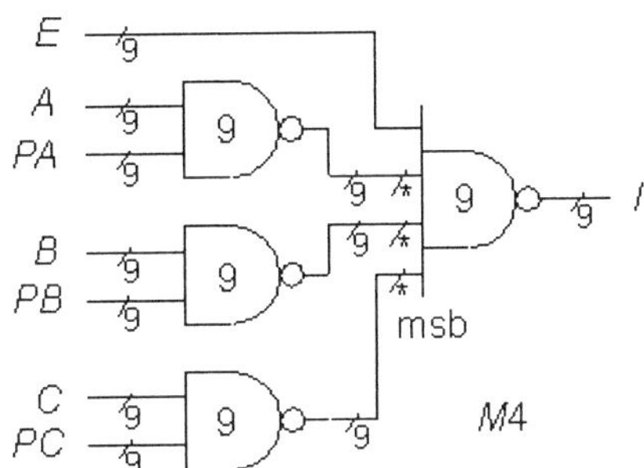

Module M5

Module M5 is a 9-line-to-4-line priority encoder. An inverter was added to output '421gat' to form Chan[3] for a correct truth table.

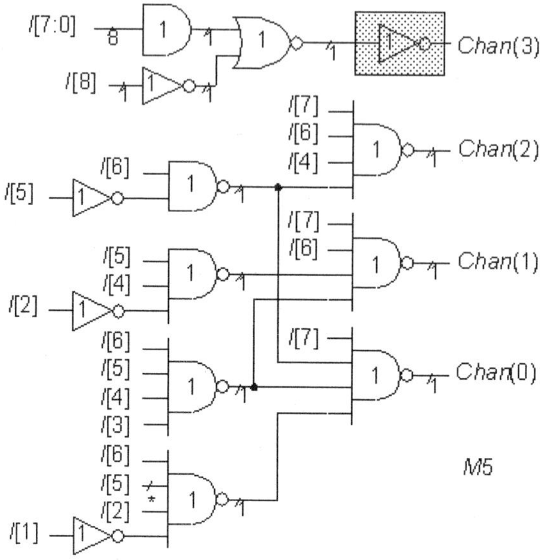

Bus Functions

I/O Bus	Function
A[8:0]	Highest priority input bus
B[8:0]	Middle priority input bus
C[8:0]	Lowest priority input bus
E[8:0]	Channel enable input bus
PA,PB,PC	Requesting bus output
Chan[3:0]	Requesting channel output

C499 / C1355

32-Bit Single-Error-Correcting Circuit

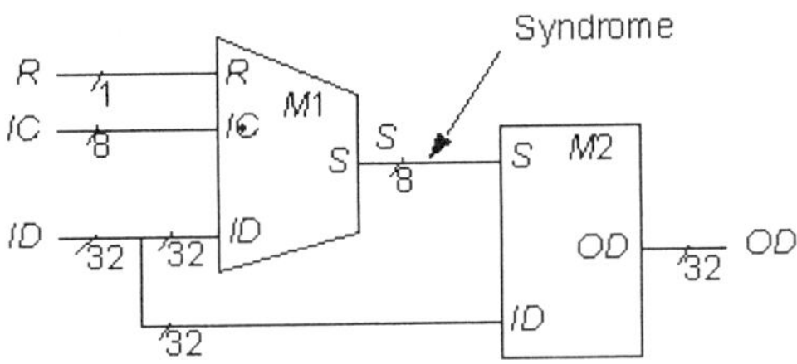

Statistics: 41 inputs; 32 outputs; 202/546 gates

Function: *c499* was found to be a single-error-correcting circuit as shown above. The 41 inputs are combined to form an 8-bit internal bus *S*, which then combines with 32 primary inputs to form the 32 primary outputs. The Boolean expressions defining *S* form the H matrix for a (40,32) Hamming code [See C. L. Chen and M. Y. Hsiao. *Error-Correcting Codes for Semiconductor Memory Applications: A State-of-the-Art Review*. IBM Journal of Research & Development, vol. 28, pp. 124-134, March 1984]. An example *S* bit is given by $S_0 = (ID_{00} \oplus ID_{04} \oplus ID_{08} \oplus ID_{12}) \oplus (ID_{16} \oplus ID_{17} \oplus ID_{18} \oplus ID_{19}) \oplus (ID_{20} \oplus ID_{21} \oplus ID_{22} \oplus ID_{23}) \oplus R \cdot IC_0$. Hence, it is in module *M1* that most XOR gates lay.

Module *M2* contains the necessary correcting logic, so *c499* can correct single-bit errors; however, no error-detection logic is present. The *S* lines are formulated to generate a unique syndrome for each input line in error. The syndromes are the column vectors of H. If syndrome *i* is seen, output OD_i is inverted. This is specified by the 32 output equations realized by *M2*. As an example, $OD_{00} = S_0 \overline{S_1 S_2 S_3} \cdot S_4 \overline{S_5 S_6 S_7} \oplus ID_{00}$.

The *c1355* circuit has the same overall function as *c499*; it differs in that all XOR primitives of *c499* are expanded to their four-NAND-gate equivalents.

Bus Functions

I/O buses	Function
ID[0:31]	Input data
IC[0:7]	Input code
R	Read line
OD[0:31]	Corrected output data

C880
8-Bit ALU

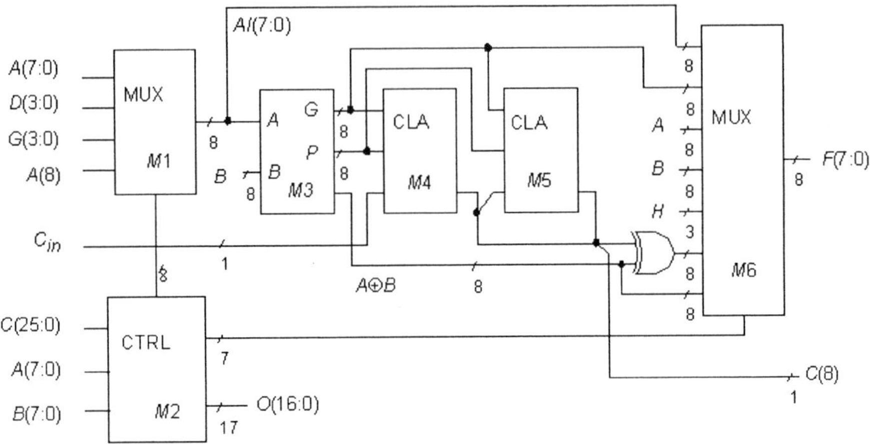

Statistics: 60 inputs; 26 outputs; 383 gates

Function: *c880* is an 8-bit ALU with the high-level model shown in above. Given the presence of a CLA module in the 74181 ALU, it is not surprising to find a similar module in *c880*. The core of this 8-bit ALU is an 8-bit 74283-style adder. The multiplexers *M1* and *M6* are both controlled by module *M2* in a fashion reminiscent of horizontal microcode; i.e., an external source must ensure that no more than one function is activated at a time on *C(25:0)*.

Bus Functions

I/O buses	Function
A[8:0]	Main A bus
B[7:0]	Main B bus
C[25:0]	Control bus
D[3:0]	4-bit bus
F[7:0]	Output function
G[3:0]	4-bit bus
C in	Carry in
C8	Carry out

C1908
16-bit error detector/corrector

High-Level View of c1980

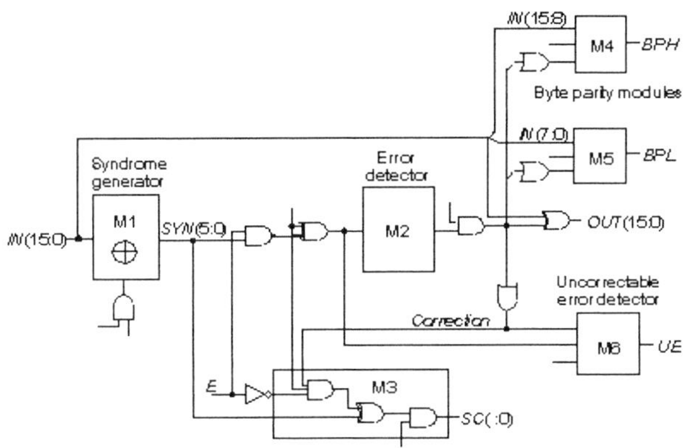

Statistics: 33 inputs; 25 outputs; 880 gates

Function: This is a 16-bit single-error-correcting and double-error-detecting (SEC/DED) circuit with some byte-error detection capability. It generates a 6-bit syndrome from the 16-bit data input *IN*, which is decoded to find the bit in error, if any. If an error is detected and the control inputs are set appropriately, error correction is performed. *C1908* has an output indicating an uncorrectable error; this is set when more than one erroneous bit is detected. The circuit can also generate syndrome bits, which are sent out via the *SC* lines. The external syndrome lines make it possible to cascade several copies of *c1908* so that detection and correction can be done for words of size greater than 16. This circuit is quite similar to the Advanced Micro Devices Am2960 16-bit error detection and correction unit.

Detailed Circuit Diagram of c1980

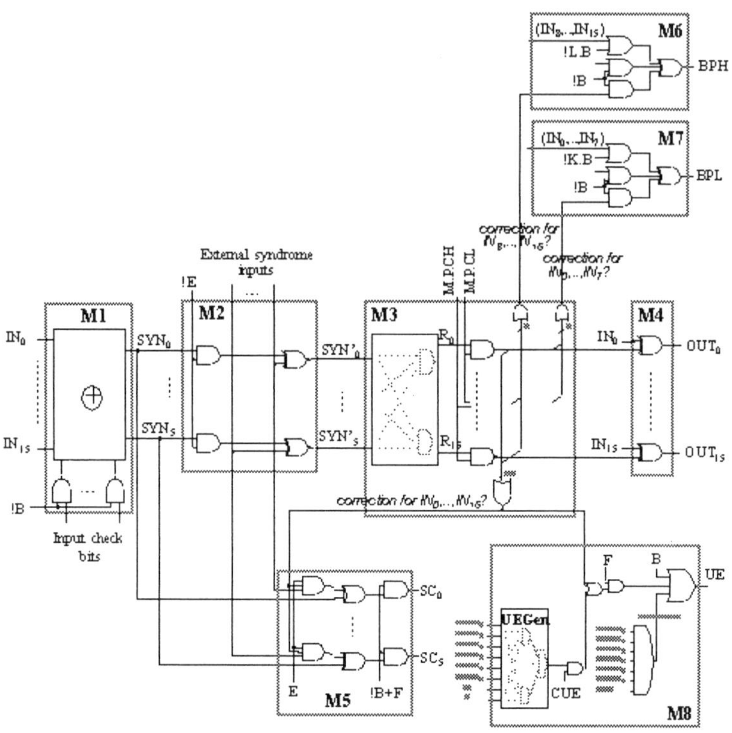

Modules

M1: generates a 6-bit syndrome (*SYN[5:0]*)
M2: may modify the syndrome with external inputs; outputs *SYN'[5:0]*
M3: decodes the syndrome to identify the bit in error, if any. Consists of 16 AND gates.
M4: corrects the input bit in error
M5: produces the output syndrome *SC[5:0]*
M6-M7: calculate a parity bit for the high (M6) and low byte (M7)
M8: asserts its output *UE* if an uncorrectable error is found in the input data bus

Inputs/Outputs

Inputs	Outputs
InDataBus[15:0] (IN)	OutDataBus[15:0] (OUT)
InCheckBits[5:0]	OutSynCheckBits[5:0] (SC)
InExtSynBits[3:0]	ByteParHi (BPH)
E,B,F (control inputs)	ByteParLo (BPL)
G,H,K,L (control inputs)	UncorrError (UE)

C2670
12-bit ALU and controller

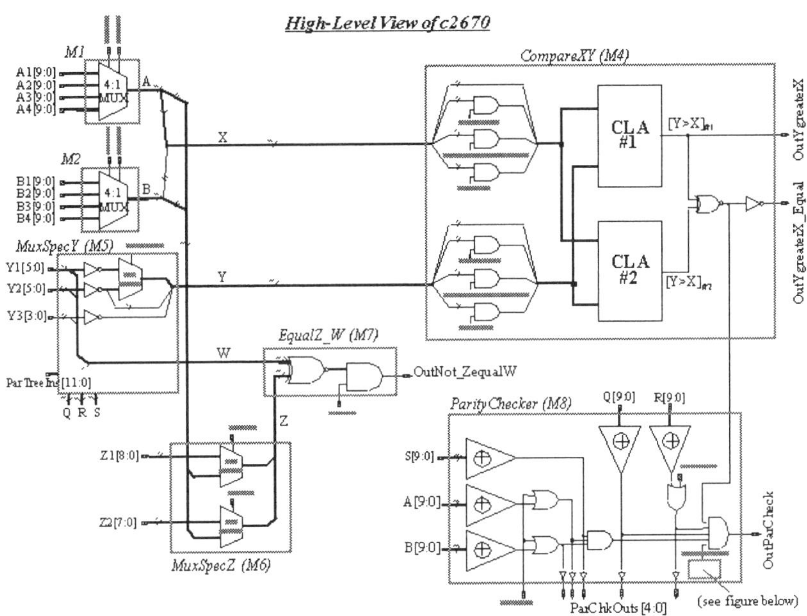

Statistics: 233 inputs; 140 outputs; 1193 gates

Function: This benchmark consists of an ALU with a comparator, an equality checker, and several parity trees. The comparator has two 12-bit inputs X and Y, and computes $Y > X$ using a carry-lookahead adder (CLA) that performs the addition $!X + Y$. It can be programmed to do a 4, 6, 8 or 12-bit comparison of its inputs.

Module *M7* (*EqualZ_W*) performs an equality check on two 17-bit buses. The *ParityChecker* module (*M8*) contains five 10-input parity trees, whose outputs are all ANDed. This module seems to perform a sanity check on the input buses of *c2670*. There are also several small pieces of logic which are mostly random.

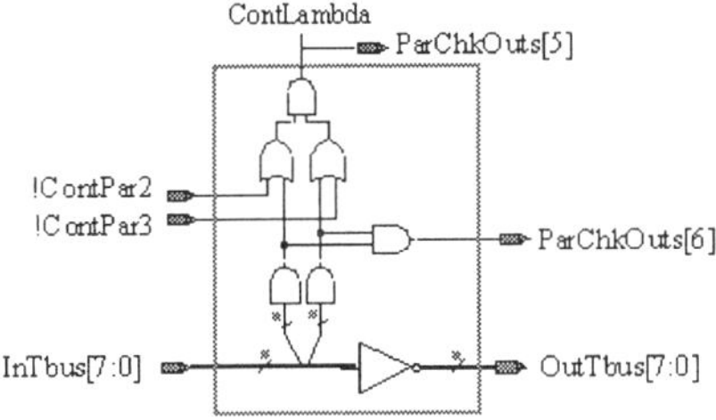

Inputs/Outputs

Inputs
input buses A[9:0], B[9:0]
Input buses Y1[5:0], Y2[5:0], Y3[3:0]
Inputs X[11:0], Y[11:0] of CompareXY (M4)
Inputs W[16:0], Z[16:0] of Bitwise Comparator (M7)
Q[9:0], R[9:0], S[9:0] (inputs to ParityChecker M8) *
ContA0, ContA1
ContB0, ContB1
ContZ0. ContZ1
ContEq
ContMask0,1,2
ContAlpha
ContBeta
ContPar0,1,2,3
ParTreeIns[11:0] (fanin of Q,R,S)
InTbus[7:0]
MiscRandomIns[11:0]
MiscMuxIn
MiscMuxCont0,1

Outputs
OutYgreaterX (Y>X)
OutYgreaterX_Equal $(Y>X)_{\#1} = (Y>X)_{\#2}$
OutZequalW (Z==W)
OutNot_ZequalW (Z != W)
OutParCheck
OutNot_ParCheck
ParChkOuts[7:0]
OutTbus[7:0]
MiscMuxOuts[10:0]
MiscBusOuts[12:0]
MiscRandomOuts[17:0]

* *Q[9:0]*, *R[9:0]* and *S[9:0]* are multiplexed out of *Y1*, *Y2*, *Y3* and *ParTreeIns*

C3540
8-bit ALU with arithmetic, logic and shift operations

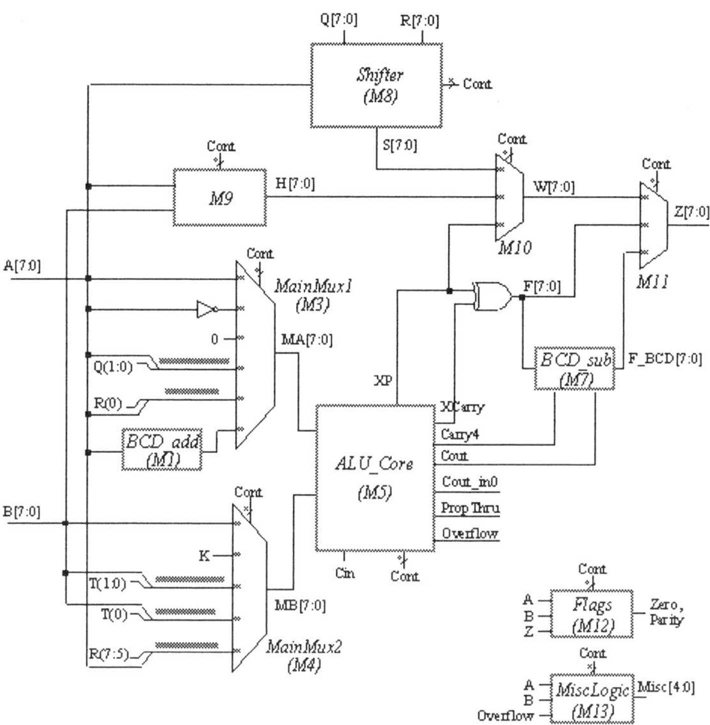

Statistics: 50 inputs; 22 outputs; 1669 gates

Function: This benchmark is an 8-bit ALU that can perform binary and BCD arithmetic operations as well as logic and shift operations. BCD addition is done via a two's-complement adder by adding 6 to both digits of the first operand, and then subtracting 6 from the digits of the result if they do not generate a carry. A total of 14 control inputs are used for multiplexing and masking data inputs. The largest module is *M5* (*ALU_Core*), which consists of two 4-bit CLAs. Module *M8* (Shifter) can shift the input bus *A* by 1 to 8 bits in either direction. Parity and zero flags are generated by module *M12* (Flags) using the input buses *A*, *B* and the output bus *Z*. Various logic functions of *A* and *B* are calculated by module *M13* which does not have an apparent high-level structure.

Modules M1 (BCD_add) and M7 (BCD_sub)

In order to perform BCD addition with a two's complement adder, module *M1* adds 6 to each digit of the input bus, and module M7 subtracts 6 from each digit of the result if there is no carry from that digit.

Module M3 (MainMux1)

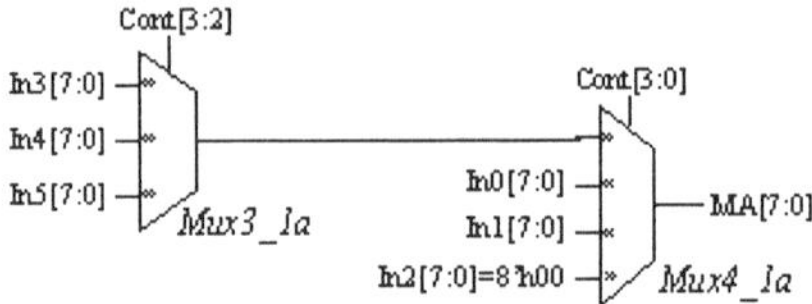

This module consists of two cascaded multiplexers, as shown above. The control signals *Cont[3:0]* determine the select inputs of *M3*.

Module M4 (MainMux2)

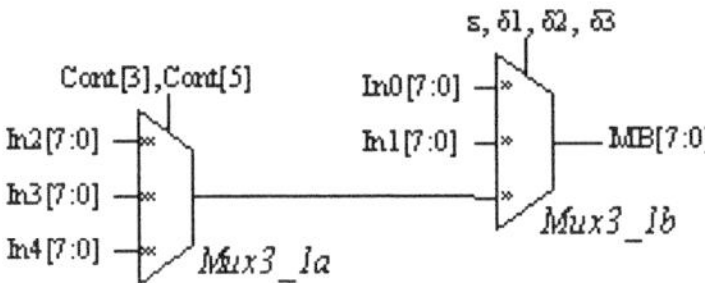

Like *M3*, Module *M4* is made up of two cascaded multiplexers. However, its select logic is more involved. The control signals *Cont[6:3]* and *Cont[1:0]* are decoded into the select signals *CHi*, *CLo1*, *CLo2* and *CLo3*.

Module M5 (ALU_Core)

This is the largest module of the *c3540* benchmark. The control inputs to this module are *Cont[12:7]* and *Cont[2:0]*. An internal signal called *Mode* determines whether a logic or an arithmetic operation is to be performed. *Mode* is 1 for a logic operation, and 0 for an arithmetic operation. In the case of arithmetic operations, an additional control signal named *Mask7_6* is used to mask bits #7 and #6 of the *MA* bus.

A block named *Logic_and_GP* computes both logic operations as well as the generate and propagate signals used for binary addition. The carry signals are computed by *CalcCarry*, and the final result of the ALU is obtained by XORing the carry signals with a modified propagate signal called *XP*.

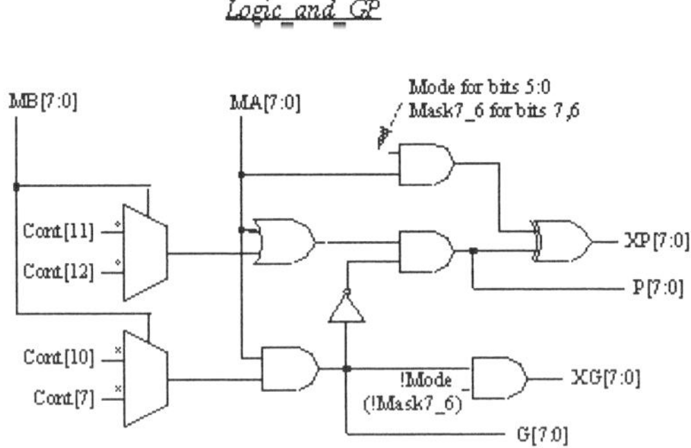

Calc Carry

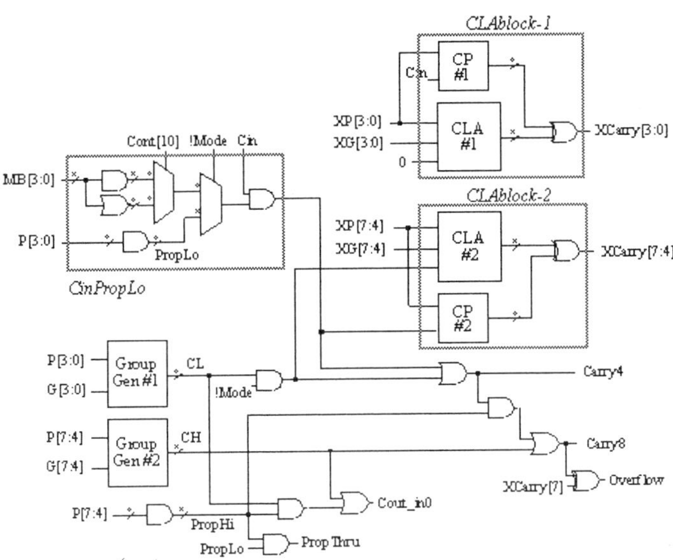

Carry Propagate (CP)

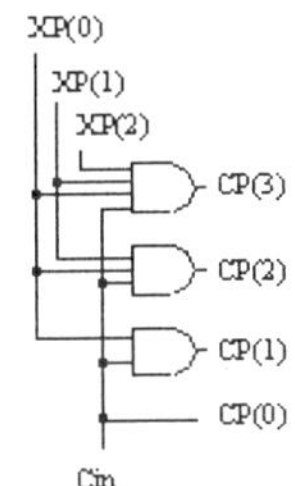

Module M8 (Shifter)

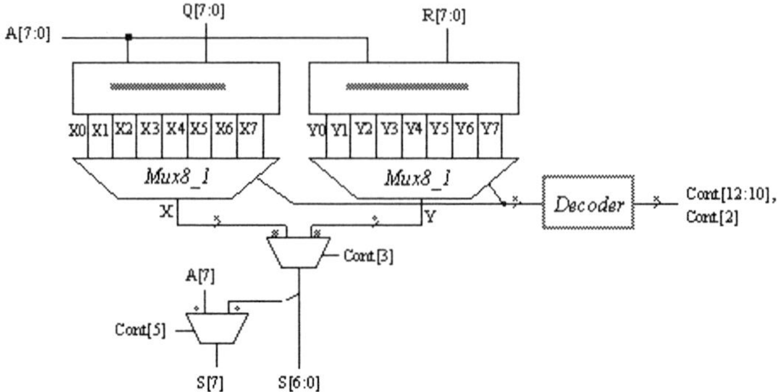

This module contains logic for shifting the input bus A by 1 to 8 bits in either direction. When shifting towards LSB (MSB), the empty bit positions are filled by a Q (R) bus. The Shifter decodes the control signals $Cont[12:10]$, and $Cont[2]$ into 8 signals, which are the select inputs for sixteen 8:1 multiplexers.

The X and Y buses are fed into a 2:1 multiplexer controlled by $Cont[3]$. There is an additional multiplexer for bit #7 whose second input is $A[7]$; this can be used for shifting signed input data.

Module M9

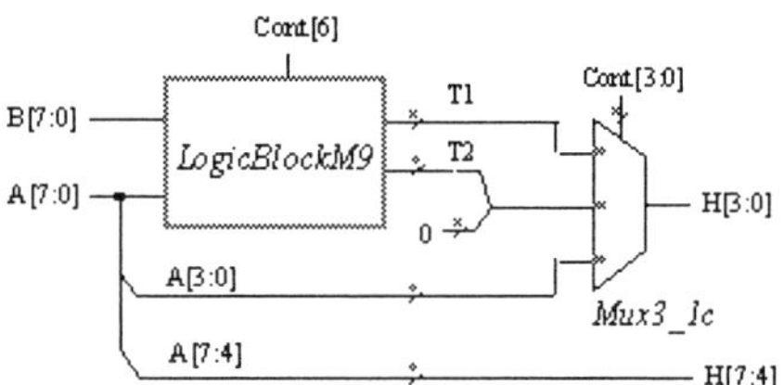

This is a relatively small module that appears to calculate some special-purpose logic functions of the input buses A and B. Five control inputs to this module are $Cont[6]$ and $Cont[3:0]$.

The details of *LogicBlockM9* are shown below. It seems to calculate a non-standard function. It may be some type of code translation or encryption.

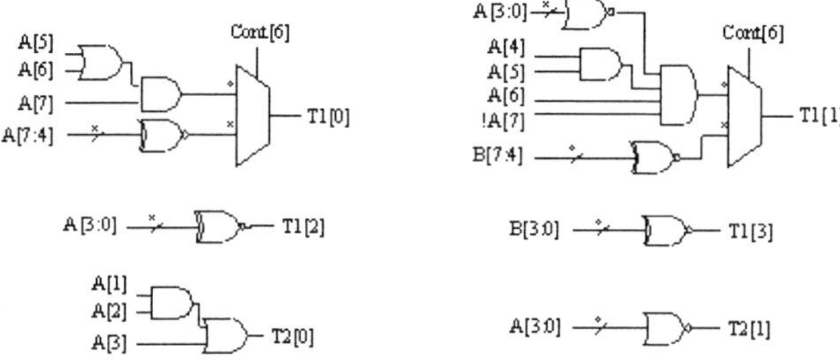

Modules M10 and M11

Both these modules contain a set of eight 3:1 multiplexers of type *Mux3_1c*.

Module M12 (Flags)

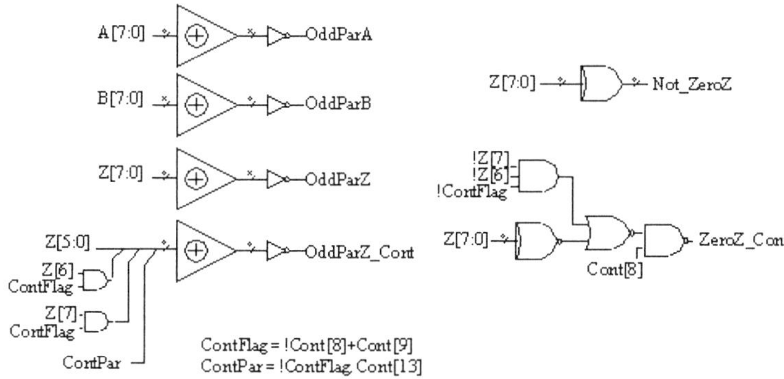

This module generates two zero flags and four parity outputs. As shown above, the parities are calculated from the input buses *A* and *B*, and the output bus *Z*. The zero flags are calculated from the output bus *Z* only. Three control inputs to *M12* are *Cont[13]* and *Cont[9:8]*.

Inputs/Outputs

Input	Output
A[7:0]	Z[7:0]
B[7:0]	OddParZ, OddParZ_Cont
Q[7:0]	OddParA
R[7:0)	OddParB
T[1:0]	NotZeroZ, ZeroZ_Cont
Cin	XCarry2, Cout_in0
K	PropThru
Cont[13:0]	MiscOuts[4:0]

C5315

9-bit ALU

Statistics: 178 inputs; 123 outputs; 2406 gates

Function: This benchmark is an ALU that performs arithmetic and logic operations simultaneously on two 9-bit input data words, and also computes the parity of the results. Modules *M6* and *M7* each compute an arithmetic or logic operation specified by the control input bus *CF[7:0]*. Module *M5* consists of multiplexers that route the results of *M6* and *M7* and four input buses to its four outputs. Output buses *OF1* and *OF2* can also be set to **logic 0** by *MuxSel[8]*. Modules *M3* and *M4* compute the parity of the result of the operation given by *CP=CF[7:4]*. Module *M5* contains four multiplexers which direct the parity results and an additional set of four inputs to its outputs. The adders in *M6* and *M7* as well as the parity logic for the arithmetic operations in *M3* and *M4* use a carry-select scheme with 4-bit (low-order) and 5-bit (high-order) blocks. The circuit also includes logic for calculating various zero and parity flags of the input buses.

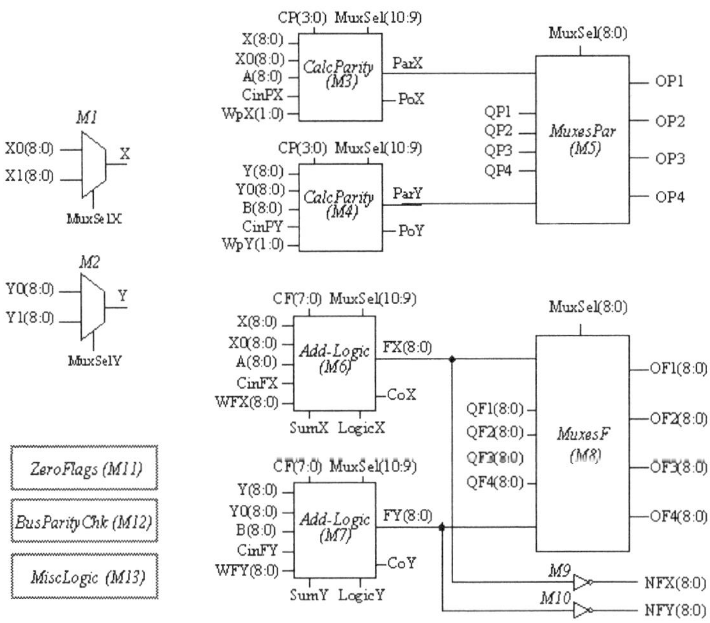

High-Level Model of c5315 [con't]

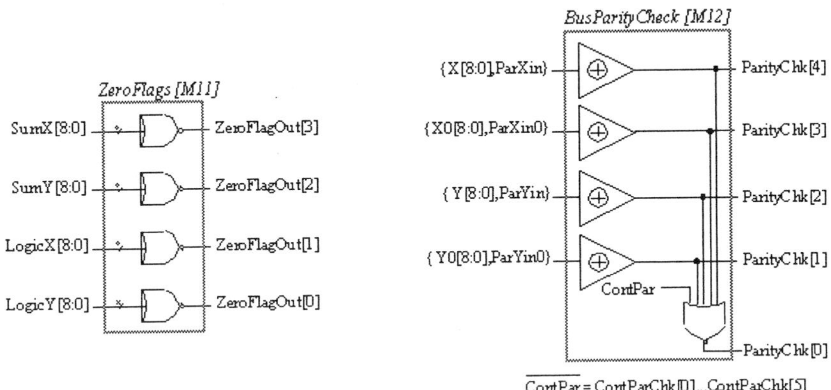

High-Level Model of c5315 [con't]

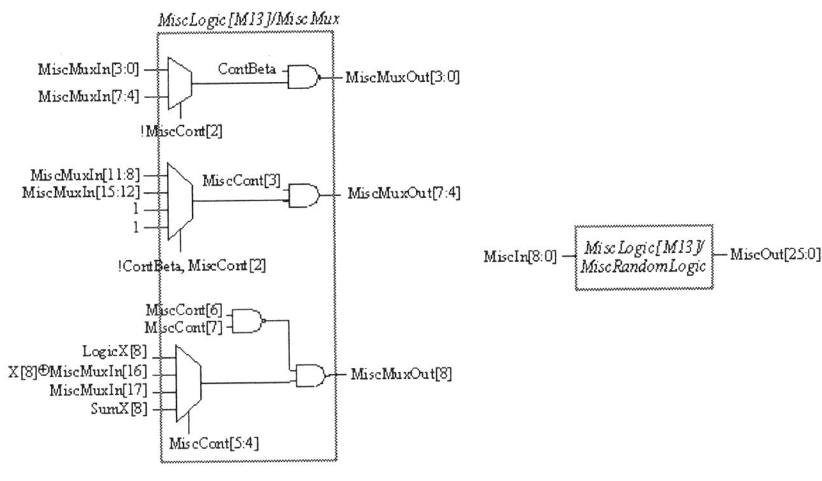

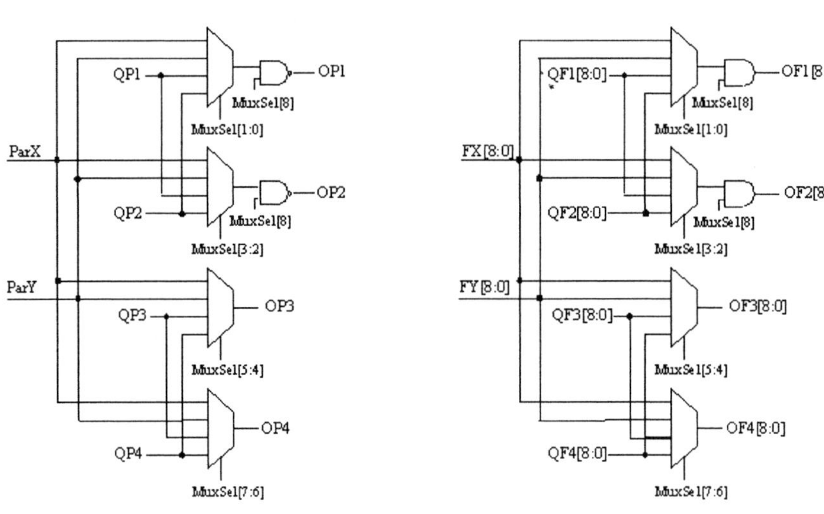
MuxesPar [M5]
QP1
OP1
MuxSel[8]
MuxSel[1:0]
ParX
QP2
OP2
MuxSel[8]
MuxSel[3:2]
ParY
QP3
OP3
MuxSel[5:4]
OP4
QP4
MuxSel[7:6]
MuxesF [M8]
QF1[8:0]
OF1[8:0]
MuxSel[8]
MuxSel[1:0]
FX[8:0]
QF2[8:0]
OF2[8:0]
MuxSel[8]
MuxSel[3:2]
FY[8:0]
QF3[8:0]
OF3[8:0]
MuxSel[5:4]
OF4[8:0]
QF4[8:0]
MuxSel[7:6]

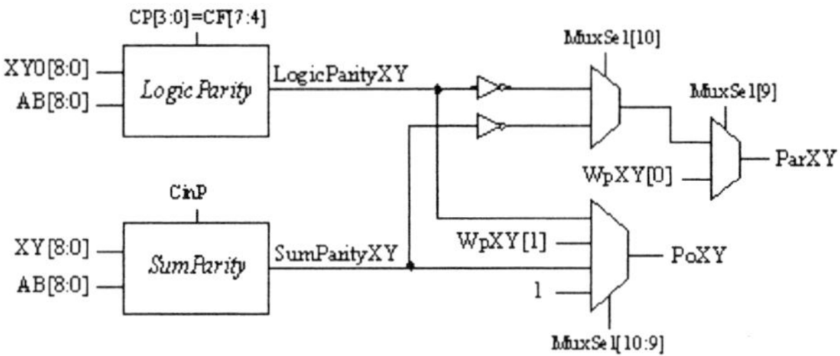
Parity [M3-M4]
CP[3:0]=CF[7:4]
XY0[8:0]
AB[8:0]
Logic Parity
LogicParityXY
MuxSel[10]
MuxSel[9]
ParXY
WpXY[0]
CinP
XY[8:0]
AB[8:0]
SumParity
SumParityXY
WpXY[1]
PoXY
1
MuxSel[10:9]

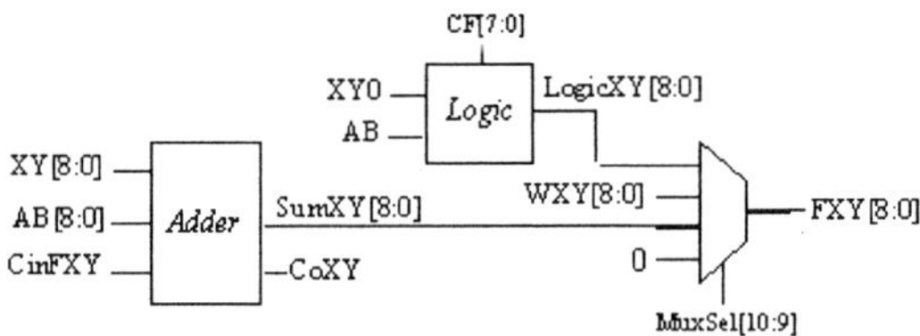
Add-Logic [M6-M7]
CF[7:0]
XY0
AB
Logic
LogicXY[8:0]
XY[8:0]
AB[8:0]
CinFXY
Adder
SumXY[8:0]
CoXY
WXY[8:0]
FXY[8:0]
0
MuxSel[10:9]

M3/Logic Parity

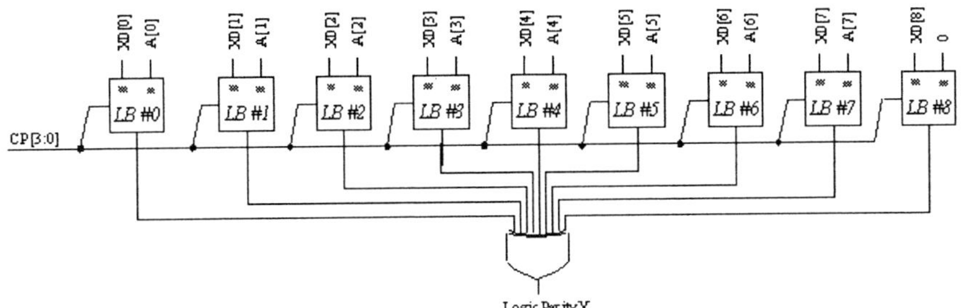

M4/Logic Parity

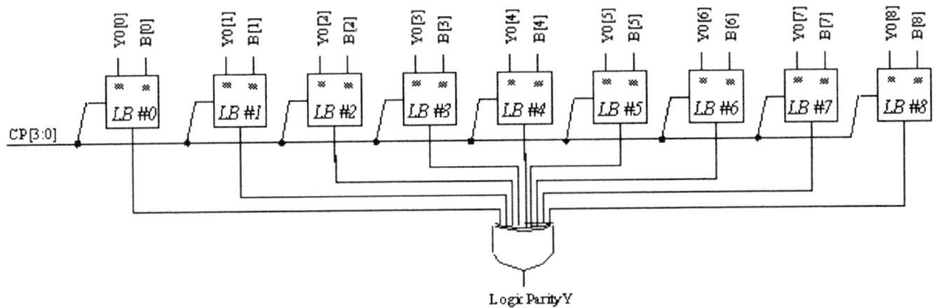

Logic Block [LB]

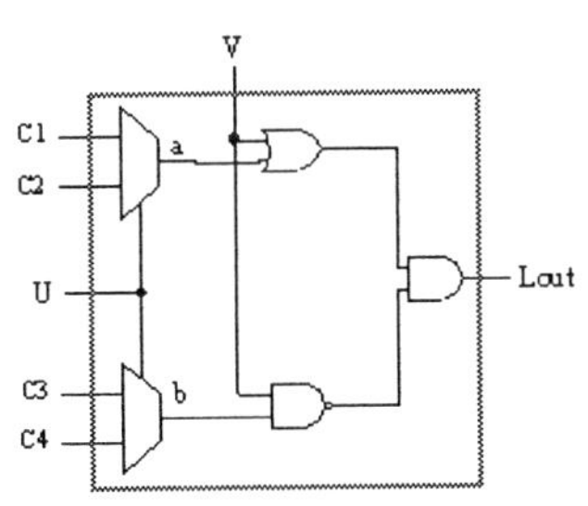

C1	C2	C3	C4	a	b	Lout
0	0	0	0	0	0	V
0	0	0	1	0	U	U'.V
0	0	1	0	0	U'	U.V
0	0	1	1	0	1	0
0	1	0	0	U	0	U+V
0	1	0	1	U	U	U⊕V
0	1	1	0	U	U'	U
0	1	1	1	U	1	U.V'
1	0	0	0	U'	0	U'+V
1	0	0	1	U'	U	U'
1	0	1	0	U'	U'	[U⊕V]'
1	0	1	1	U'	1	U'.V'
1	1	0	0	1	0	1
1	1	0	1	1	U	U'+V'
1	1	1	0	1	U'	U+V'
1	1	1	1	1	1	V'

* Lout: all the 16 functions of two variables (U, V)

SumParity

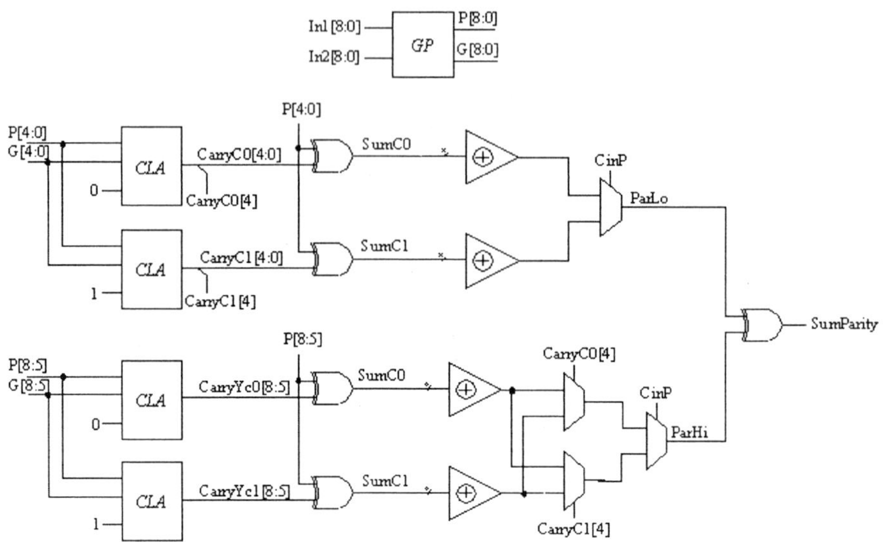

Note: Carry[i] = carry from bit position i to i+1.

M6/Logic

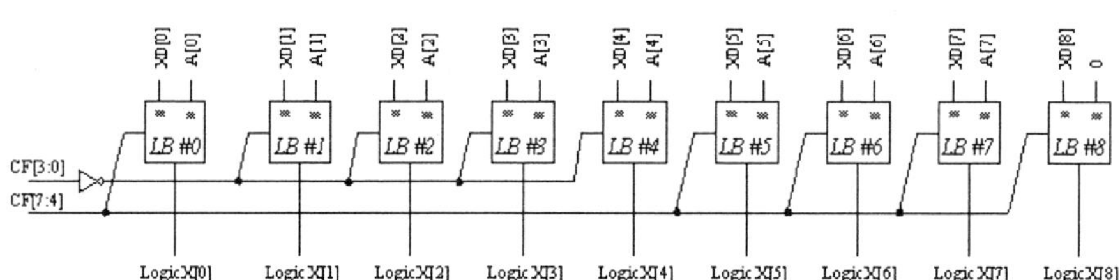

M7/Logic

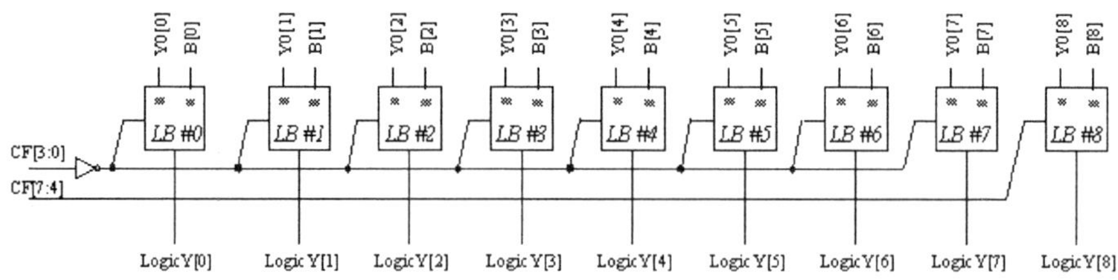

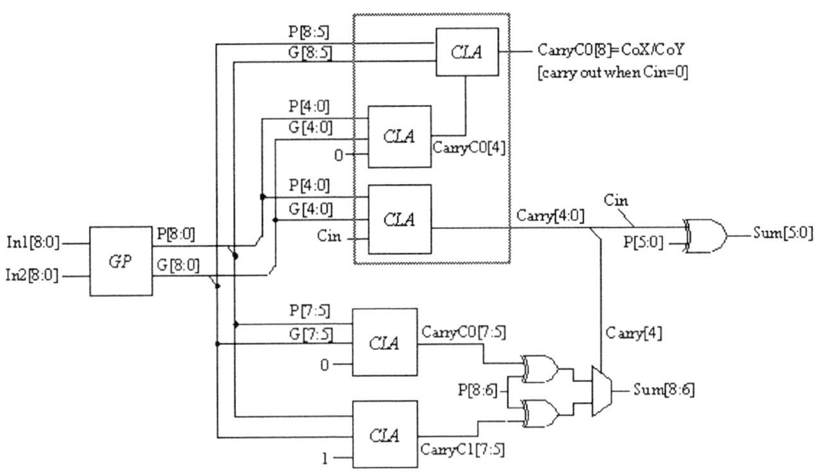

Inputs/Outputs

Input	Output
X0[8:0], X1[8:0]	OP1,OP2,OP3,OP4
MuxSelX	OF1[8:0]
A[8:0]	OF2[8:0]
Y0[8:0], Y1[8:0]	OF3[8:0]
MuxSelY	OF4[8:0]
B[8:0]	NXF[8:0]
CinFX, CinFY	NYF[8:0]
CinPX, CinPY	CoX,CoY
WpX[1:0]	PoX,PoY
WpY[1:0]	ParityChk[4:0]
QP1,QP2,QP3,QP4	ZeroFlagOut[3:0]
Q1[8:0], Q2[8:0], Q3[8:0], Q4[8:0]	MiscMuxOut[10:0]
WFX[8:0]	MiscOut[25:0]
WFY[8:0]	
MuxSel[10:0]	
CF[7:0]	
CP[3:0]=CF[7:4]	
ParYin= MuxSelY ? ParYin0 : ParYin1	
(ParYin0, ParYin1)	
ParXin= MuxSelX ? ParXin0 : ParXin1	
(ParXin0, ParXin1)	
ContParChk[5:0]	
MiscMuxIn[17:0]	
MiscContIn[7:0]	
MiscIn[8:0]	

C6288
16-bit multiplier

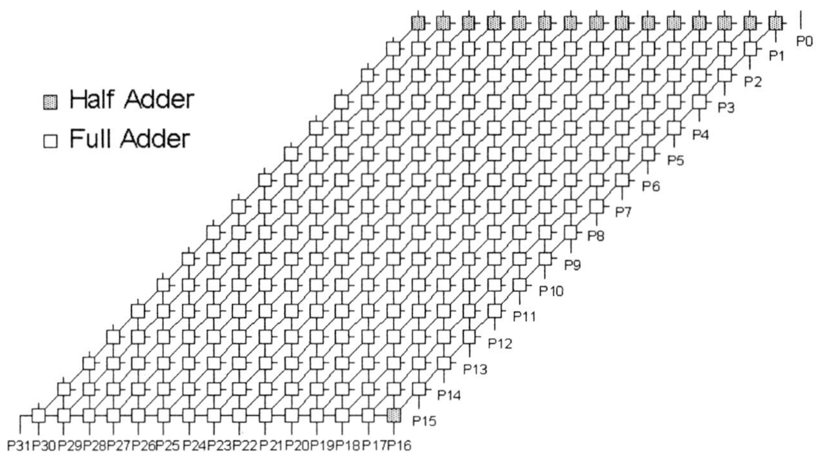

Statistics: 32 inputs; 32 outputs; 2406 gates

Function: The *c6288* benchmark, whose multiplication function was previously known, represents a much larger gate-level circuit that also has a concise functional description. The figure above shows how the 2406 gates form 240 full and half adder cells arranged in a 15x16 matrix. An alternate representation is shown below, and the adder cells are detailed below that.

Alternate Depiction

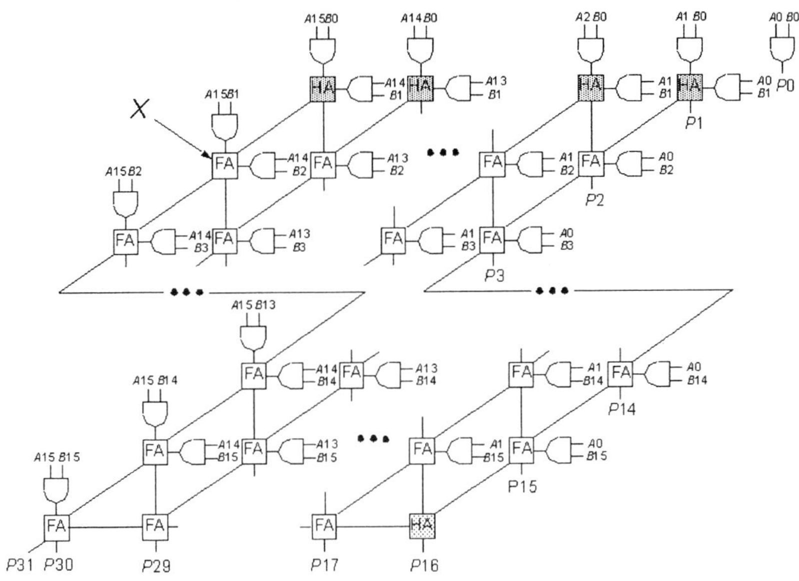

Full Adder Module

The 15 top-row half adders lack the C_i input; each has two inverters at locations V. The single half adder in the bottom row lacks the B input, thereby acquiring two inverters at locations W.

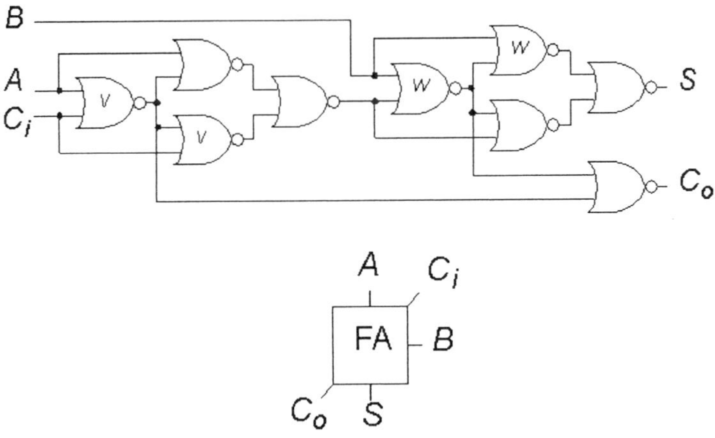

Bus Functions

I/O Busses	Function
A[15:0]	A bus
B[15:0]	B bus
P[31:0]	Product bus

C7552
34-bit adder, comparator and parity checker

High-Level Model of c7552

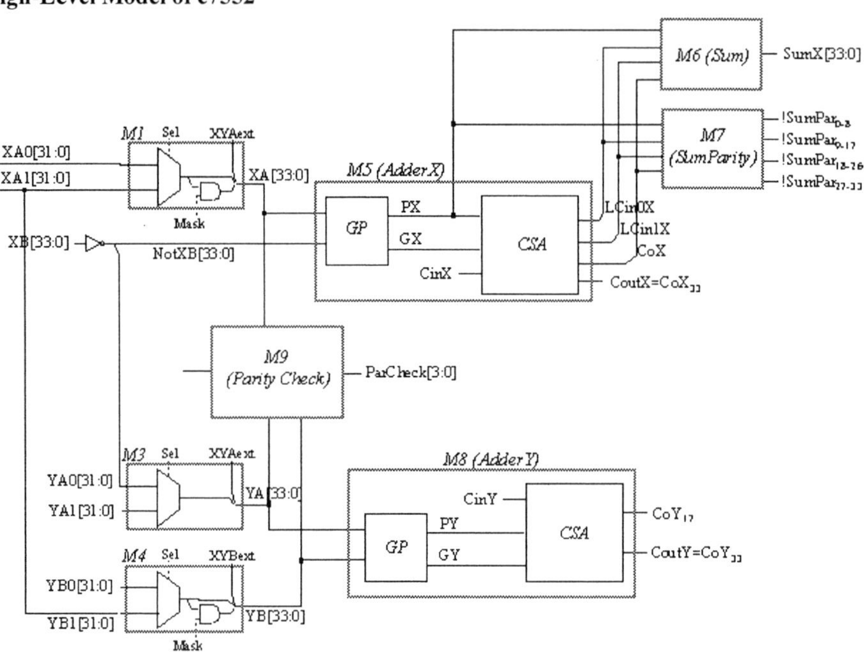

Statistics: 207 inputs; 108 outputs; 3512 gates

Function: This benchmark circuit contains a 34-bit adder (*M5*), a 34-bit magnitude comparator (*M8*) using another 34-bit adder, and a parity checker (*M9*). Each of the *XA*, *YA*, and *YB* buses is fed by a set of 2:1 multiplexers controlled by the *Sel* input. Bits 31-22 of *XA* and *YB* can be set to **logic 0** with the *Mask* input. The two adders *M5* and *M8* are identical, and are of carry select type, as are those of *c5315*. They consist of alternating 4- and 5-bit blocks, with the last block being 2 bits. The comparator (*M8*) of this benchmark is similar to that of *c2670*. It performs the comparison *YB>XB* (if *Sel*=0) or *YB>!YA1* (if *Sel*=1) by calculating *YB+!XB* (if *Sel*=0) or *YB+!YA1* (if *Sel*=1) (Note: the input bus *YA1* is assumed to be inverted). The comparator has an output (*CoutY*) for the whole 34-bit inputs as well as an output (*CoutY_17*) for the 17-bit portion of its inputs. Module *M7* calculates the parity for the following four parts of the adder output *SumX*: *SumX[8:0]*, *SumX[17:9]*, *SumX[26:18]*, *SumX[33:27]*. Module *M9* appears to be a type of sanity checker that calculates the AND of the parities of all its inputs.

M5 / CSA

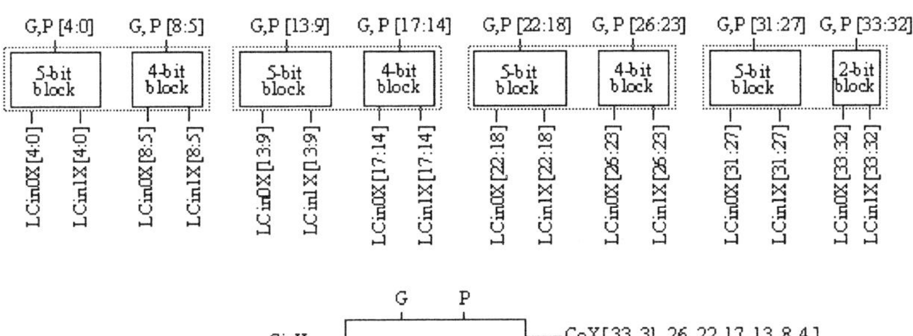

Note: Carry[i] = carry from bit position i to (i+1)

Module M6 (SUM)

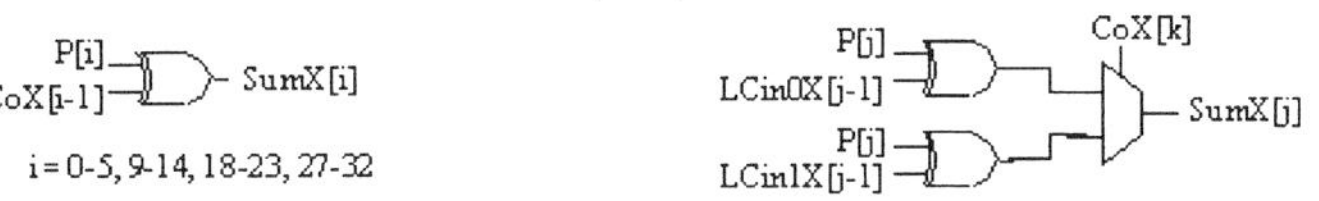

i = 0-5, 9-14, 18-23, 27-32

j = 6-8, 15-17, 24-26, 33, k = pivot carries: 4, 13, 22, 31

Module M7 (SUM PARITY)

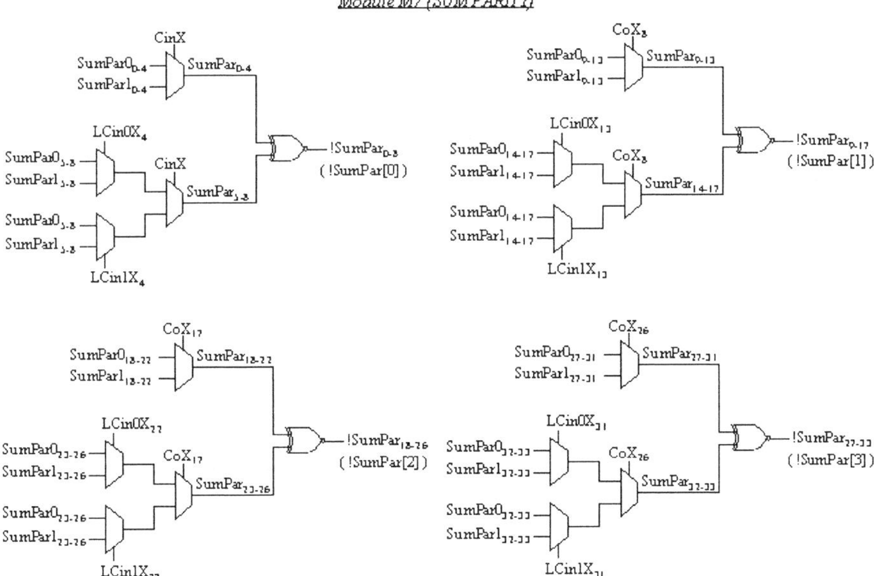

Module M9 (PARITY CHECK)

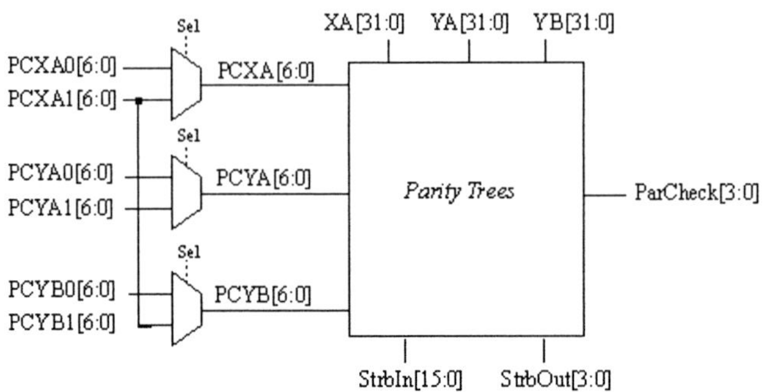

M9 / Parity Trees

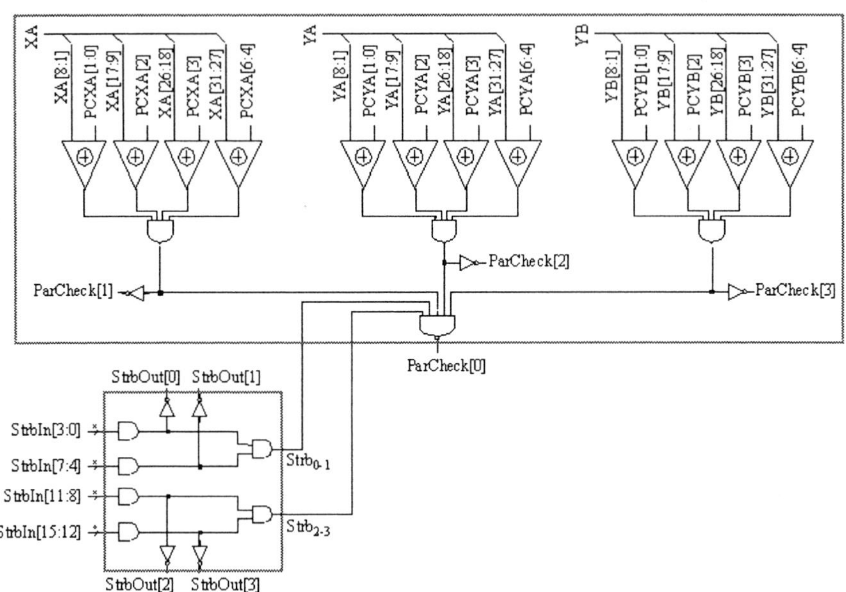

Inputs/Outputs

Input	Output
XA0[31:0]	SumX[33:0]
XA1[31:0]	!SumPar[3:0]
XB[33:0]	CoutX1, CoutX2
YA1[31:0]	CoutY1, CoutY2
YB0[33:0]	CoutY_17
!Sel	ParCheck[3:0]
CinX, CinY	XBbuf[33:0]
XYAext, XYBext	StrbOut
PCXA0[6:0]	PCYA0buf[3:0]
PCXA1[6:0]	MiscOut[5:0]
PCYA0[6:0]	
PCYA1[6:0]	
PCYB0[6:0]	
Mask=!Mask1+!Mask2	
Mask1, Mask2	
StrbIn[15:0]	
MiscIn[7:0]	

Appendix B

— Logics —

Logics

In this appendix we present the 'truth' tables for all operations (including S-buffers) of logics covered in this thesis, namely Boolean logic, 3-, 4-, 5-, 8- and 9-valued logics, extended logic, local search logic, logic over Booleans and sets, and sets logic. While the first six logics described have a known and constant number of values (2, 3, 4, 5, 8 and 9), the other four have an arbitrary number of values dependent either on the number of primary inputs (PIs) of the agent, or on the number of possible diagnoses/theories considered in the problem.

For each logic we start by displaying the number of theories encoded, the ability to handle unspecified values, the purpose of the logic and its values and meanings; then we show the logic tables for the logic operations.

Boolean Logic

Theories	1
Unspecifiedness	no

Purpose: Boolean simulation

Values
0
1

Operations

NOT	
0	1
1	0

AND	0	1
0	0	0
1	0	1

OR	0	1
0	0	1
1	1	1

XOR	0	1
0	0	1
1	1	0

S-Buffer s-a-0	
0	0
1	0

S-Buffer s-a-1	
0	1
1	1

3-valued Logic

Theories	1
Unspecifiedness	√

Purpose: simulation

Values	meaning
0	constant value 0
1	constant value 1
x	unspecified / unknown

Operations

NOT	
0	1
1	0
x	x

AND	**0**	**1**	x
0	0	0	0
1	0	1	x
x	0	x	x

OR	**0**	**1**	x
0	0	1	x
1	1	1	1
x	x	1	x

XOR	**0**	**1**	x
0	0	1	x
1	1	0	x
x	x	x	x

S-Buffer s-a-0	
0	0
1	0
x	0

S-Buffer s-a-1	
0	1
1	1
x	1

4-valued Logic

Theories	2
Unspecifiedness	no

Purpose: Test Generation (TG) or Differential Test Generation (DTG)

Values	meaning (for TG)	meaning (for DTG)
0	0 in normal and faulty theories	0 in both faulty theories
1	1 in normal and faulty theories	1 in both faulty theories
D	1 in normal theory; 0 in faulty theory	1 in 1st faulty theory; 0 in 2nd faulty theory
$\overline{D}$	0 in normal theory; 1 in faulty theory	0 in 1st faulty theory; 1 in 2nd faulty theory

Operations

NOT	
0	1
1	0
D	$\overline{D}$
$\overline{D}$	D

AND	**0**	**1**	**D**	**$\overline{D}$**
0	0	0	0	0
1	0	1	D	$\overline{D}$
D	0	D	D	0
$\overline{D}$	0	$\overline{D}$	0	$\overline{D}$

OR	**0**	**1**	**D**	**$\overline{D}$**
0	0	1	D	$\overline{D}$
1	1	1	1	1
D	D	1	D	1
$\overline{D}$	$\overline{D}$	1	1	$\overline{D}$

XOR	**0**	**1**	**D**	**$\overline{D}$**
0	0	1	D	$\overline{D}$
1	1	0	$\overline{D}$	D
D	D	$\overline{D}$	0	1
$\overline{D}$	$\overline{D}$	D	1	0

Input	S-buffer/0 output	S-buffer/1 output
$0 = 0/0$	$0/0 = 0$	$0/1 = \overline{D}$
$1 = 1/1$	$1/0 = D$	$1/1 = 1$
$D = 1/0$	$1/0 = D$	$1/1 = 1$
$\overline{D} = 0/1$	$0/0 = 0$	$0/1 = \overline{D}$

5-valued Logic

Theories	2
Unspecifiedness	√

Purpose: Test Generation

Values	Meaning
0	0 in normal and faulty theories
1	1 in normal and faulty theories
D	1 in normal theory; 0 in faulty theory
$\overline{D}$	0 in normal theory; 1 in faulty theory
x	unspecified/unknown in either the normal or faulty theory

Operations

NOT	
0	1
1	0
D	$\overline{D}$
$\overline{D}$	D
x	x

AND	0	1	D	$\overline{D}$	x
0	0	0	0	0	0
1	0	1	D	$\overline{D}$	x
D	0	D	D	0	x
$\overline{D}$	0	$\overline{D}$	0	$\overline{D}$	x
x	0	x	x	x	x

OR	0	1	D	$\overline{D}$	x
0	0	1	D	$\overline{D}$	x
1	1	1	1	1	1
D	D	1	D	1	x
$\overline{D}$	$\overline{D}$	1	1	$\overline{D}$	x
x	x	1	x	x	x

XOR	0	1	D	$\overline{D}$	x
0	0	1	D	$\overline{D}$	x
1	1	0	$\overline{D}$	D	x
D	D	$\overline{D}$	0	1	x
$\overline{D}$	$\overline{D}$	D	1	0	x
x	x	x	x	x	x

Input	S-buffer/0 output	S-buffer/1 output
0	0	$\overline{D}$
1	D	1
D	D	1
$\overline{D}$	0	$\overline{D}$
x	x	x

8-valued Logic

Theories	3
Unspecifiedness	no

Purpose: Differential Test Generation (diagnosis, fault location)

Values	meaning
0	0 in normal and faulty theories
1	1 in normal and faulty theories
d_1-0	0 except in 1st faulty theory
d_1-1	1 except in 1st faulty theory
d_2-0	0 except in 2nd faulty theory
d_2-1	1 except in 2nd faulty theory
m-0	0 in normal theory; 1 in either faulty theory
m-1	1 in normal theory; 0 in either faulty theory

Denoting the 1st and 2nd faulty theories by, respectively, F_1 (with set of faults ϕ_1) and F_2 (with set of faults ϕ_2), the meaning of the 8 values, for any $X \in \{0,1\}$, is thus the following:

X the truth value is independent from faults ϕ_1 and ϕ_2. These faults have no influence on the physical truth-value which is always X.

d_1-X the truth value depends on faults ϕ_1 but not on ϕ_2. In model F_1, where ϕ_1 occurs, the physical truth-value is the complement of X; otherwise (in models N and F_2), it is X.

d_2-X similar to the previous, with ϕ_1 and ϕ_2 swapping roles.

m-X the truth value depends on both diagnoses ϕ_1 and ϕ_2. In both models F_1 and F_2, the physical truth-value is the complement of X, obtained in model N.

In general, the encoded truth-value Z of some signal line in this logic is shown in the table below (where N denotes the fault-free theory), for all possible combinations of Boolean values in the 3 theories:

N	0	0	0	0	1	1	1	1
F_1	0	0	1	1	0	0	1	1
F_2	0	1	0	1	0	1	0	1
Z	0	d_2-0	d_1-0	m-0	m-1	d_1-1	d_2-1	1

Operations

NOT	0	1	d_1-0	d_1-1	d_2-0	d_2-1	m-0	m-1
	1	0	d_1-1	d_1-0	d_2-1	d_2-0	m-1	m-0

AND	m-0	d_2-0	d_1-0	0	1	d_1-1	d_2-1	m-1
m-0	m-0	d_2-0	d_1-0	0	m-0	d_2-0	d_1-0	0
d_2-0	d_2-0	d_2-0	0	0	d_2-0	d_2-0	0	0
d_1-0	d_1-0	0	d_1-0	0	d_1-0	0	d_1-0	0
0	0	0	0	0	0	0	0	0
1	m-0	d_2-0	d_1-0	0	1	d_1-1	d_2-1	m-1
d_1-1	d_2-0	d_2-0	0	0	d_1-1	d_1-1	m-1	m-1
d_2-1	d_1-0	0	d_1-0	0	d_2-1	m-1	d_2-1	m-1
m-1	0	0	0	0	m-1	m-1	m-1	m-1

OR	m-0	d_2-0	d_1-0	0	1	d_1-1	d_2-1	m-1
m-0	m-0	m-0	m-0	m-0	1	1	1	1
d_2-0	m-0	d_2-0	m-0	d_2-0	1	d_1-1	1	d_1-1
d_1-0	m-0	m-0	d_1-0	d_1-0	1	1	d_2-1	d_2-1
0	m-0	d_2-0	d_1-0	0	1	d_1-1	d_2-1	m-1
1	1	1	1	1	1	1	1	1
d_1-1	1	d_1-1	1	d_1-1	1	d_1-1	1	d_1-1
d_2-1	1	1	d_2-1	d_2-1	1	1	d_2-1	d_2-1
m-1	1	d_1-1	d_2-1	m-1	1	d_1-1	d_2-1	m-1

XOR	m-0	d_2-0	d_1-0	0	1	d_1-1	d_2-1	m-1
m-0	0	d_1-0	d_2-0	m-0	m-1	d_2-1	d_1-1	1
d_2-0	d_1-0	0	m-0	d_2-0	d_2-1	m-1	1	d_1-1
d_1-0	d_2-0	m-0	0	d_1-0	d_1-1	1	m-1	d_2-1
0	m-0	d_2-0	d_1-0	0	1	d_1-1	d_2-1	m-1
1	m-1	d_2-1	d_1-1	1	0	d_1-0	d_2-0	m-0
d_1-1	d_2-1	m-1	1	d_1-1	d_1-0	0	m-0	d_2-0
d_2-1	d_1-1	1	m-1	d_2-1	d_2-0	m-0	0	d_1-0
m-1	1	d_1-1	d_2-1	m-1	m-0	d_2-0	d_1-0	0

An S-buffer can be so because it belongs to set of faults ϕ_1 or ϕ_2 or to both. There are 8 possibilities of dependency of an S-buffer to consider, as in the table below, where F_i / B as dependency means that the gate is stuck-at-B in diagnosis ϕ_i (theory F_i):

Dependency \ Input	m-0	d_2-0	d_1-0	0	1	d_1-1	d_2-1	m-1
$F_1/0$, $F_2/0$	0	0	0	0	m-1	m-1	m-1	m-1
$F_2/0$	d_1-0	0	d_1-0	0	d_2-1	m-1	d_2-1	m-1
$F_1/0$	d_2-0	d_2-0	0	0	d_1-1	d_1-1	m-1	m-1
$F_1/0$, $F_2/1$	d_2-0	d_2-0	d_2-0	d_2-0	d_1-1	d_1-1	d_1-1	d_1-1
$F_1/1$, $F_2/0$	d_1-0	d_1-0	d_1-0	d_1-0	d_2-1	d_2-1	d_2-1	d_2-1
$F_1/1$	m-0	m-0	d_1-0	d_1-0	1	1	d_2-1	d_2-1
$F_2/1$	m-0	d_2-0	m-0	d_2-0	1	d_1-1	1	d_1-1
$F_1/1$, $F_2/1$	m-0	m-0	m-0	m-0	1	1	1	1

9-valued Logic

Theories	2
Unspecifiedness	√

Purpose: Test Generation

Values	Meaning
0/0	0 in normal and faulty theories
1/1	1 in normal and faulty theories
u/u	unspecified/unknown in both theories
0/1	0 in normal theory; 1 in faulty theory
1/0	1 in normal theory; 0 in faulty theory
u/0	unspecified/unknown in normal theory; 0 in faulty theory
u/1	unspecified/unknown in normal theory; 1 in faulty theory
0/u	0 in normal theory; unspecified/unknown in faulty theory
1/u	1 in normal theory; unspecified/unknown in faulty theory

Operations

NOT	0/0	0/1	0/u	1/0	1/1	1/u	u/0	u/1	u/u
	1/1	1/0	1/u	0/1	0/0	0/u	u/1	u/0	u/u

AND	0/0	0/1	0/u	1/0	1/1	1/u	u/0	u/1	u/u
0/0	0/0	0/0	0/0	0/0	0/0	0/0	0/0	0/0	0/0
0/1	0/0	0/1	0/u	0/0	0/1	0/u	0/0	0/1	0/u
0/u	0/0	0/u	0/u	0/0	0/u	0/u	0/0	0/u	0/u
1/0	0/0	0/0	0/0	1/0	1/0	1/0	u/0	u/0	u/0
1/1	0/0	0/1	0/u	1/0	1/1	1/u	u/0	u/1	u/u
1/u	0/0	0/u	0/u	1/0	1/u	1/u	u/0	u/u	u/u
u/0	0/0	0/0	0/0	u/0	u/0	u/0	u/0	u/0	u/0
u/1	0/0	0/1	0/u	u/0	u/1	u/u	u/0	u/1	u/u
u/u	0/0	0/u	0/u	u/0	u/u	u/u	u/0	u/u	u/u

OR	0/0	0/1	0/u	1/0	1/1	1/u	u/0	u/1	u/u
0/0	0/0	0/1	0/u	1/0	1/1	1/u	u/0	u/1	u/u
0/1	0/1	0/1	0/1	1/1	1/1	1/1	u/1	u/1	u/1
0/u	0/u	0/1	0/u	1/u	1/1	1/u	u/u	u/1	u/u
1/0	1/0	1/1	1/u	1/0	1/1	1/u	1/0	1/1	1/u
1/1	1/1	1/1	1/1	1/1	1/1	1/1	1/1	1/1	1/1
1/u	1/u	1/1	1/u	1/u	1/1	1/u	1/u	1/1	1/u
u/0	u/0	u/1	u/u	1/0	1/1	1/u	u/0	u/1	u/u
u/1	u/1	u/1	u/1	1/1	1/1	1/1	u/1	u/1	u/1
u/u	u/u	u/1	u/u	1/u	1/1	1/u	u/u	u/1	u/u

XOR	0/0	0/1	0/u	1/0	1/1	1/u	u/0	u/1	u/u
0/0	0/0	0/1	0/u	1/0	1/1	1/u	u/0	u/1	u/u
0/1	0/1	0/0	0/u	1/1	1/0	1/u	u/1	u/0	u/u
0/u	0/u	0/u	0/u	1/u	1/u	1/u	u/u	u/u	u/u
1/0	1/0	1/1	1/u	0/0	0/1	0/u	u/0	u/1	u/u
1/1	1/1	1/0	1/u	0/1	0/0	0/u	u/1	u/0	u/u
1/u	1/u	1/u	1/u	0/u	0/u	0/u	u/u	u/u	u/u
u/0	u/0	u/1	u/u	u/0	u/1	u/u	u/0	u/1	u/u
u/1	u/1	u/0	u/u	u/1	u/0	u/u	u/1	u/0	u/u
u/u	u/u	u/u	u/u	u/u	u/u	u/u	u/u	u/u	u/u

Input	S-buffer/0 output	S-buffer/1 output
0/0	0/0	0/1
0/1	0/0	0/1
0/u	0/0	0/1
1/0	1/0	1/1
1/1	1/0	1/1
1/u	1/0	1/1
u/0	u/0	u/1
u/1	u/0	u/1
u/u	u/0	u/1

Extended Logic

Theories	1
Unspecifiedness	√

Purpose: Simulation

Values	meaning
0	constant value 0
1	constant value 1
id-0	unspecified/unknown value from source *id* with inversion parity 0
id-1	unspecified/unknown value from source *id* with inversion parity 1

(Shaded lines denote values that are not necessarily unique. Rather, they represent an arbitrary number of values, since symbols shown in italic stand for any possible instantiation of the general entity they represent, as the *id* source in this case, standing for any possible signal line.)

Operations

NOT	**0**	**1**	***id*-0**	***id*-1**
	1	0	*id*-1	*id*-0

$G=X_1 \wedge X_2$	**0**	**1**	$id_1\text{-}0$	$id_1\text{-}1$	$id_2\text{-}0$	$id_2\text{-}1$
0	0	0	0	0	0	0
1	0	1	id_1-0	id_1-1	id_2-0	id_2-1
id_1**-0**	0	id_1-0	id_1-0	0	id_G-0	id_G-0
id_1**-1**	0	id_1-1	0	id_1-1	id_G-0	id_G-0
id_2**-0**	0	id_2-0	id_G-0	id_G-0	id_2-0	0
id_2**-1**	0	id_2-1	id_G-0	id_G-0	0	id_2-1

(id_1 and id_2 are used to denote different sources.
id_G stands for the identification of the gate.)

$G=X_1 \vee X_2$	**0**	**1**	$id_1\text{-}0$	$id_1\text{-}1$	$id_2\text{-}0$	$id_2\text{-}1$
0	0	1	id_1-0	id_1-1	id_2-0	id_2-1
1	1	1	1	1	1	1
id_1**-0**	id_1-0	1	id_1-0	1	id_G-0	id_G-0
id_1**-1**	id_1-1	1	1	id_1-1	id_G-0	id_G-0
id_2**-0**	id_2-0	1	id_G-0	id_G-0	id_2-0	1
id_2**-1**	id_2-1	1	id_G-0	id_G-0	1	id_2-1

$G=X_1\oplus X_2$	0	1	id_1-0	id_1-1	id_2-0	id_2-1
0	0	1	id_1-0	id_1-1	id_2-0	id_2-1
1	1	0	id_1-1	id_1-0	id_2-1	id_2-0
id_1-**0**	id_1-0	id_1-1	0	1	id_G-0	id_G-0
id_1-**1**	id_1-1	id_1-0	1	0	id_G-0	id_G-0
id_2-**0**	id_2-0	id_2-1	id_G-0	id_G-0	0	1
id_2-**1**	id_2-1	id_2-0	id_G-0	id_G-0	1	0

S-Buffer s-a-0 input	output
0	0
1	0
id-**0**	0
id-**1**	0

S-Buffer s-a-1 input	output
0	1
1	1
id-**0**	1
id-**1**	1

Local Search Logic

Theories	#PI + 1
Unspecifiedness	$\sqrt{}$

Purpose: Local search for maximisation of unspecified inputs in test patterns

Values	meaning
Set : 0	originally 0
Set : 1	originally 1
Set : *id*-0	originally unspecified/unknown value from source *id* with inversion parity 0
Set : *id*-1	originally unspecified/unknown value from source *id* with inversion parity 1

Set stands for a set of conditional values in the form $id_{PI} / Value_{PI}$ meaning that if PI id_{PI} alone becomes unspecified, then the real 'circuit' signal s assumes a *different* value $Value_{PI}$ (an unspecified value of the extended logic), i.e. denoting the logic value of signal s after unspecification of PI i by $s \propto i$, then $s \propto id_{PI} = Value_{PI}$.

Thus, *Set* precises the cases when the signal value differs from the original upon 'unspecification' of a PI. Hence, each local search logic value encodes the extended logic value for a specific input vector t, together with the extended logic values when each specified PI of t is turned into unspecified.

The general rule for computing a gate output in this logic is to find all values for the relevant input vectors (the original one and the ones with a PI turned into unspecified). This procedure is exemplified with the *and*-operation in the pseudo-code below:

```
Procedure And (In: Sx:Vx,Sy:Vy, Out: Sz:Vz);
    Vz ← Vx and Vy      % compute original output value.
    Sz0 ← {}            % start with empty set.
    for each i such that i/_ ∈ Sx ∪ Sy do   % conditional values.
        i/VXi ∈ Sx or else Vxi ← Vx    % x ∝ i
        i/VYi ∈ Sy or else Vyi ← Vy    % y ∝ i
        VZi ← VXi and VYi              % z ∝ i
        if VZi ≠ Vz then Sz0 ← Sz0 ∪ {i/VZi}
    end for
    Sz ← Sz0
end Procedure
```

Operations

NOT	
S:0	$\{i/id\text{-}0: i/id\text{-}1 \in S\} \cup \{i/id\text{-}1: i/id\text{-}0 \in S\}$:1
S:1	$\{i/id\text{-}0: i/id\text{-}1 \in S\} \cup \{i/id\text{-}1: i/id\text{-}0 \in S\}$:0
$S : id$-0	$\{i/id\text{-}0: i/id\text{-}1 \in S\} \cup \{i/id\text{-}1: i/id\text{-}0 \in S\}$: id-1
$S : id$-1	$\{i/id\text{-}0: i/id\text{-}1 \in S\} \cup \{i/id\text{-}1: i/id\text{-}0 \in S\}$: id-0

In the following tables, $P \in \{0,1\}$, $X \oplus Y = X \cup Y \setminus X \cap Y = X \setminus Y \cup Y \setminus X$, and $\{id_X, id_Y\}$ stand for different sources.

Each output logic value Z of the form '*Set$_Z$: original Z*' is shown by displaying resulting *Set$_Z$* and *original Z* in separate columns. (Original Z is just the result of performing the logic operation over original arguments in the extended logic shown previously.)

The cases considered depend on the original values of arguments X and Y. Their sets of conditional values are denoted as S_X and S_Y, respectively.

original X	original Y	Set$_Z$ of (Z = X *and* Y)	original Z
0	0	$\{i/id\text{-}P: i/id\text{-}P \in S_X \cap S_Y\} \cup$ $\{i/z\text{-}0: i/id_X\text{-}P_X \in S_X \wedge i/id_Y\text{-}P_Y \in S_Y\}$	0
1	1	$\{i/id\text{-}P: i/id\text{-}P \in S_X \wedge \neg\exists_{idy,Py}\, i/id_Y\text{-}P_Y \in S_Y\} \cup$ $\{i/id\text{-}P: i/id\text{-}P \in S_Y \wedge \neg\exists_{idx,Px}\, i/id_X\text{-}P_X \in S_X\} \cup$ $\{i/z\text{-}0: i/id_X\text{-}P_X \in S_X \wedge i/id_Y\text{-}P_Y \in S_Y\}$	1
0	1	$\{i/id\text{-}P: i/id\text{-}P \in S_X \wedge \neg\exists_{idy,Py}\, i/id_Y\text{-}P_Y \in S_Y\} \cup$ $\{i/z\text{-}0: i/id_X\text{-}P_X \in S_X \wedge i/id_Y\text{-}P_Y \in S_Y\}$	0
0	$id_Y\text{-}P_Y$	$\{i/id\text{-}P: i/id\text{-}P \in S_X \cap S_Y\} \cup$ $\{i/id_Y\text{-}P_Y: i/id_Y\text{-}P_Y \in S_X \wedge \neg\exists_{id,P}\, i/id\text{-}P \in S_Y\} \cup$ $\{i/z\text{-}0: i/id_X\text{-}P_X \in S_X \setminus S_Y\}$	0
1	$id_Y\text{-}P_Y$	$\{i/id\text{-}P: i/id\text{-}P \in S_X \cap S_Y\} \cup$ $\{i/id\text{-}P: i/id\text{-}P \in S_Y \wedge \neg\exists_{idx,Px}\, i/id_X\text{-}P_X \in S_X\} \cup$ $\{i/z\text{-}0: i/id_X\text{-}P_X \in S_X \setminus S_Y\}$	$id_Y\text{-}P_Y$
$id_X\text{-}P_X$	$id_X\text{-}P_X$	$\{i/id\text{-}P: i/id\text{-}P \in S_X \cap S_Y\} \cup$ $\{i/z\text{-}0: i/id\text{-}P \in S_X \oplus S_Y\}$	$id_X\text{-}P_X$
$id_X\text{-}0$	$id_X\text{-}1$	$\{i/id\text{-}P: i/id\text{-}P \in S_X \cap S_Y\} \cup$ $\{i/z\text{-}0: i/id\text{-}P \in S_X \oplus S_Y\}$	0
$id_X\text{-}P_X$	$id_Y\text{-}P_Y$	$\{i/id\text{-}P: i/id\text{-}P \in S_X \cap S_Y\} \cup$ $\{i/id_X\text{-}P_X: i/id_X\text{-}P_X \in S_Y \wedge \neg\exists_{id,P}\, i/id\text{-}P \in S_X\} \cup$ $\{i/id_Y\text{-}P_Y: i/id_Y\text{-}P_Y \in S_X \wedge \neg\exists_{id,P}\, i/id\text{-}P \in S_Y\}$	z-0

original X	original Y	Set_Z of $(Z = X$ or $Y)$	original Z
0	0	$\{i/id\text{-}P: i/id\text{-}P \in S_X \wedge \neg\exists_{idy,Py}\ i/id_Y\text{-}P_Y \in S_Y\} \cup$ $\{i/id\text{-}P: i/id\text{-}P \in S_Y \wedge \neg\exists_{idx,Px}\ i/id_X\text{-}P_X \in S_X\} \cup$ $\{i/z\text{-}0: i/id_X\text{-}P_X \in S_X \wedge i/id_Y\text{-}P_Y \in S_Y\}$	0
1	1	$\{i/id\text{-}P: i/id\text{-}P \in S_X \cap S_Y\} \cup$ $\{i/z\text{-}0: i/id_X\text{-}P_X \in S_X \wedge i/id_Y\text{-}P_Y \in S_Y\}$	1
0	1	$\{i/id\text{-}P: i/id\text{-}P \in S_Y \wedge \neg\exists_{idx,Px}\ i/id_X\text{-}P_X \in S_X\} \cup$ $\{i/z\text{-}0: i/id_X\text{-}P_X \in S_X \wedge i/id_Y\text{-}P_Y \in S_Y\}$	1
0	$id_Y\text{-}P_Y$	$\{i/id\text{-}P: i/id\text{-}P \in S_X \cap S_Y\} \cup$ $\{i/id\text{-}P: i/id\text{-}P \in S_Y \wedge \neg\exists_{idx,Px}\ i/id_X\text{-}P_X \in S_X\} \cup$ $\{i/z\text{-}0: i/id_X\text{-}P_X \in S_X \setminus S_Y\}$	$id_Y\text{-}P_Y$
1	$id_Y\text{-}P_Y$	$\{i/id\text{-}P: i/id\text{-}P \in S_X \cap S_Y\} \cup$ $\{i/id_Y\text{-}P_Y: i/id_Y\text{-}P_Y \in S_X \wedge \neg\exists_{id,P}\ i/id\text{-}P \in S_Y\} \cup$ $\{i/z\text{-}0: i/id_X\text{-}P_X \in S_X \setminus S_Y\}$	1
$id_X\text{-}P_X$	$id_X\text{-}P_X$	$\{i/id\text{-}P: i/id\text{-}P \in S_X \cap S_Y\} \cup$ $\{i/z\text{-}0: i/id\text{-}P \in S_X \oplus S_Y\}$	$id_X\text{-}P_X$
$id_X\text{-}0$	$id_X\text{-}1$	$\{i/id\text{-}P: i/id\text{-}P \in S_X \cap S_Y\} \cup$ $\{i/z\text{-}0: i/id\text{-}P \in S_X \oplus S_Y\}$	1
$id_X\text{-}P_X$	$id_Y\text{-}P_Y$	$\{i/id\text{-}P: i/id\text{-}P \in S_X \cap S_Y\} \cup$ $\{i/id_X\text{-}P_X: i/id_X\text{-}P_X \in S_Y \wedge \neg\exists_{id,P}\ i/id\text{-}P \in S_X\} \cup$ $\{i/id_Y\text{-}P_Y: i/id_Y\text{-}P_Y \in S_X \wedge \neg\exists_{id,P}\ i/id\text{-}P \in S_Y\}$	$z\text{-}0$

X	Y	Set_Z of $(Z = X\ xor\ Y)$	original Z
0	0	$\{i/id\text{-}P: i/id\text{-}P \in S_X \wedge\neg\exists_{idy,Py}\ i/id_Y\text{-}P_Y \in S_Y\} \cup$ $\{i/id\text{-}P: i/id\text{-}P \in S_Y \wedge\neg\exists_{idx,Px}\ i/id_X\text{-}P_X \in S_X\} \cup$ $\{i/z\text{-}0: i/id_X\text{-}P_X \in S_X \wedge i/id_Y\text{-}P_Y \in S_Y\}$	0
1	1	$\{i/id\text{-}0: i/id\text{-}1 \in S_X \wedge\neg\exists_{idy,Py}\ i/id_Y\text{-}P_Y \in S_Y\} \cup$ $\{i/id\text{-}1: i/id\text{-}0 \in S_X \wedge\neg\exists_{idy,Py}\ i/id_Y\text{-}P_Y \in S_Y\} \cup$ $\{i/id\text{-}0: i/id\text{-}1 \in S_Y \wedge\neg\exists_{idx,Px}\ i/id_X\text{-}P_X \in S_X\} \cup$ $\{i/id\text{-}1: i/id\text{-}0 \in S_Y \wedge\neg\exists_{idx,Px}\ i/id_X\text{-}P_X \in S_X\} \cup$ $\{i/z\text{-}0: i/id_X\text{-}P_X \in S_X \wedge i/id_Y\text{-}P_Y \in S_Y\}$	0
0	1	$\{i/id\text{-}P: i/id\text{-}P \in S_Y \wedge\neg\exists_{idx,Px}\ i/id_X\text{-}P_X \in S_X\} \cup$ $\{i/id\text{-}0: i/id\text{-}1 \in S_X \wedge\neg\exists_{idy,Py}\ i/id_Y\text{-}P_Y \in S_Y\} \cup$ $\{i/id\text{-}1: i/id\text{-}0 \in S_X \wedge\neg\exists_{idy,Py}\ i/id_Y\text{-}P_Y \in S_Y\} \cup$ $\{i/z\text{-}0: i/id_X\text{-}P_X \in S_X \wedge i/id_Y\text{-}P_Y \in S_Y\}$	1
0	$id_Y\text{-}P_Y$	$\{i/id\text{-}P: i/id\text{-}P \in S_Y \wedge\neg\exists_{idx,Px}\ i/id_X\text{-}P_X \in S_X\} \cup$ $\{i/z\text{-}0: i/id_X\text{-}P_X \in S_X\}$	$id_Y\text{-}P_Y$
1	$id_Y\text{-}0$	$\{i/id\text{-}0: i/id\text{-}1 \in S_Y \wedge\neg\exists_{idx,Px}\ i/id_X\text{-}P_X \in S_X\} \cup$ $\{i/id\text{-}1: i/id\text{-}0 \in S_Y \wedge\neg\exists_{idx,Px}\ i/id_X\text{-}P_X \in S_X\} \cup$ $\{i/z\text{-}0: i/id_X\text{-}P_X \in S_X\}$	$id_Y\text{-}1$
1	$id_Y\text{-}1$	$\{i/id\text{-}0: i/id\text{-}1 \in S_Y \wedge\neg\exists_{idx,Px}\ i/id_X\text{-}P_X \in S_X\} \cup$ $\{i/id\text{-}1: i/id\text{-}0 \in S_Y \wedge\neg\exists_{idx,Px}\ i/id_X\text{-}P_X \in S_X\} \cup$ $\{i/z\text{-}0: i/id_X\text{-}P_X \in S_X\}$	$id_Y\text{-}0$
$id_X\text{-}P_X$	$id_X\text{-}P_X$	$\{i/z\text{-}0: i/id\text{-}P \in S_X \oplus S_Y\}$	0
$id_X\text{-}0$	$id_X\text{-}1$	$\{i/z\text{-}0: i/id\text{-}P \in S_X \wedge\neg\exists_{idy,Py}\ i/id_Y\text{-}P_Y \in S_Y\} \cup$ $\{i/z\text{-}0: i/id_Y\text{-}P_Y \in S_Y \wedge\neg\exists_{id,P}\ i/id\text{-}P \in S_X\} \cup$ $\{i/z\text{-}0: i/id\text{-}P \in S_X \wedge i/id_Y\text{-}P_Y \in S_Y\}$	1
$id_X\text{-}P_X$	$id_Y\text{-}P_Y$	$\{\}$	z-0

S-Buffer s-a-0 input	output	S-Buffer s-a-1 input	output
S:0	{}:0	*S*:0	{}:1
S:1	{}:0	*S*:1	{}:1
S : *id*-0	{}:0	*S* : *id*-0	{}:1
S : *id*-1	{}:0	*S* : *id*-1	{}:1

Logic over Booleans and Sets

Theories	1 + number of possible diagnoses
Unspecifiedness	no

Purpose: Simulation, diagnosis

Values	meaning
Set - 0	0 in the fault-free theory; 1 for any diagnosis of *Set*
Set - 1	1 in the fault-free theory; 0 for any diagnosis of *Set*

Set is thus the set of diagnoses that inverts the normal value.

Operations

NOT	
L-0	L-1
L-1	L-0

AND	L_Y-0	L_Y-1
L_X-0	$L_X \cap L_Y$ -0	$L_X \setminus L_Y$ - 0
L_X-1	$L_Y \setminus L_X$ - 0	$L_X \cup L_Y$ -1

OR	L_Y-0	L_Y-1
L_X-0	$L_X \cup L_Y$ - 0	$L_Y \setminus L_X$ - 1
L_X-1	$L_X \setminus L_Y$ - 1	$L_X \cap L_Y$ - 1

XOR	L_Y-0	L_Y-1
L_X-0	$L_X \oplus L_Y$ - 0	$L_X \oplus L_Y$ - 1
L_X-1	$L_Y \oplus L_X$ - 1	$L_X \oplus L_Y$ - 0

S-Buffer	
L_i-0	$L_{S1} \cup (L_i \setminus L_{S0})$ - 0
L_i-1	$L_{S0} \cup (L_i \setminus L_{S1})$ - 1

Where $X \oplus Y = X \cup Y \setminus X \cap Y = X \setminus Y \cup Y \setminus X$.

S-buffers are placed at the output of gates suspected to be faulty (in the universe of the possible diagnoses D). So, an S-buffer for gate g has associated a set of the possible diagnoses (diagnostic sets) where g appears as stuck. We call this set L_S. Since g can appear either as stuck-at-0 or stuck-at-1, we split this set in two (L_{S0} and L_{S1}), one for each type of fault, as below:

$L_{S0} = \{Diag \in D:\ g/0 \in Diag\}$

$L_{S1} = \{Diag \in D:\ g/1 \in Diag\}$

$L_S = L_{S0} \cup L_{S1}$ (of course, $L_{S0} \cap L_{S1} = \varnothing$)

Sets Logic

Theories	number of possible diagnoses
Unspecifiedness	no

Purpose: General diagnostic problems

Values	Meaning
Set	1 for the diagnoses in *Set*; 0 otherwise

Set is thus the set of diagnoses (of universe D) that cause the signal to take value 1.

Operations

The logic operations are expressed by the respective set operations, where the universe is the set D of possible diagnoses:

$$not\ S = \overline{S} \qquad\qquad (\text{i.e. } not\ S = \{F \in D: F \notin S\})$$
$$X\ and\ Y = X \cap Y \qquad\qquad (\text{i.e. } X\ and\ Y = \{F: F \in X \wedge F \in Y\})$$
$$X\ or\ Y = X \cup Y \qquad\qquad (\text{i.e. } X\ or\ Y = \{F: F \in X \vee F \in Y\})$$
$$X\ xor\ Y = X \oplus Y \qquad\qquad (\text{as before, } X \oplus Y = X \cup Y \setminus X \cap Y = X \setminus Y \cup Y \setminus X)$$

$$s\_buffer(L_S, L_i) = L_{SI} \cup (L_i \setminus L_{S0})$$

We represent the output of an S-buffer for gate g with associated dependencies L_S (set of possible diagnoses where g appears as stuck), subject to input i as $s\_buffer(L_S, i)$.

As explained before, L_S is split in two (L_{S0} and L_{SI}), for both types of faults (s-a-0 and s-a-1), as below:

$$L_{S0} = \{Diag \in D: g/0 \in Diag\}$$
$$L_{SI} = \{Diag \in D: g/1 \in Diag\}$$
$$L_S = L_{S0} \cup L_{SI} \qquad (\text{of course, } L_{S0} \cap L_{SI} = \varnothing)$$

전기/전자